Functional Cobalt Oxides

Functional Cobalt Oxides

FUNDAMENTALS, PROPERTIES, AND APPLICATIONS

Tsuyoshi Takami

PAN STANFORD PUBLISHING

Published by

Pan Stanford Publishing Pte. Ltd.
Penthouse Level, Suntec Tower 3
8 Temasek Boulevard
Singapore 038988

Email: editorial@panstanford.com
Web: www.panstanford.com

British Library Cataloguing-in-Publication Data
A catalogue record for this book is available from the British Library.

Functional Cobalt Oxides: Fundamentals, Properties, and Applications

Copyright © 2014 Pan Stanford Publishing Pte. Ltd.

ISBN 978-981-4463-32-4 (Hardcover)
ISBN 978-981-4463-33-1 (eBook)

Printed in the USA

Contents

Preface   vii

Acknowledgments   ix

1 Introduction   **1**
  1.1 Prologue   1
  1.2 Charge, Spin, and Orbitals   1
  1.3 Physical Properties under External Perturbations   12

2 Spin-State Crossover   **27**
  2.1 Prologue   27
  2.2 Valence and Spin State of Co Ions   28
  2.3 Spin-State Crossover   31
  2.4 Metal-Insulator Transition   39
  2.5 Thermal Rectifier   40

3 Li Ion Battery   **45**
  3.1 Prologue   45
  3.2 History of Batteries   46
  3.3 $LiCoO_2$ Batteries   48

4 Huge Thermoelectric Power   **57**
  4.1 Prologue   57
  4.2 Thermoelectric Materials   58
  4.3 Na_xCoO_2 61
  4.4 Other Co Oxides   63
  4.5 Origin of a Huge Thermoelectric Power   64
  4.6 Application   71

5 Room-Temperature Ferromagnetism   **75**
  5.1 Prologue   75

5.2   Ferromagnetism in Transition Metal Oxides   78
5.3   Room-Temperature Ferromagnetism of $(Sr,Y)CoO_{3-\delta}$ 79

6   Partially Disordered Antiferromagnetic Transition   85
6.1   Prologue   85
6.2   Magnetism of $Ca_3Co_2O_6$ 87
6.3   Magnetism of Quasi-1D $A_{n+2}Co_{n+1}O_{3n+3}$ 92

7   Superconductivity   109
7.1   Prologue   109
7.2   Bose–Einstein Condensation   110
7.3   High-T_c Cuprate Superconductors   112
7.4   Superconductivity of $Na_xCoO_2 \cdot yH_2O$ 115

**8   Transport Properties Combined with Charge, Spin, and
     Orbital: Magnetoresistance and Spin Blockade   121**
8.1   Prologue   121
8.2   The Magnetoresistance Effect   122
8.3   Magnetoresistance of $RBaCo_2O_{5.5}$ 127
8.4   Extremely Large Magnetoresistance of $PdCoO_2$ 130
8.5   Spin Blockade   131

9   Intrinsic Inhomogeneity   137
9.1   Prologue   137
9.2   Spin Cluster   140
9.3   Polaron Cluster   142
9.4   Phase Separation   144
       9.4.1   Metallic and Semiconducting Phases   144
       9.4.2   Spin-Density Wave and Ferromagnetic Phases   148

10   Move/Diffuse and Charge/Discharge Effect   155
10.1   Prologue   155
10.2   Cathode Material of Solid Oxide Fuel Cell:
        $Sr_{0.7}Y_{0.3}CoO_{2.63}$ 157
10.3   Oxygen Storage Material: $YBaCo_4O_{7+\delta}$ 162

Index   165

Preface

When one looks around, one's apparels are found to made of many materials, which are said to reach 50,000 varieties. Surprisingly, these materials are composed of at most about 50 kinds of elements. Among them, the transition metals have a high melting point, high density, and multivalence. This book deals with oxide materials, including the transition metal Co, but the elementary Co itself with the 29th Clarke number exhibits ferromagnetism, as well as Ni and Fe, and has been used as a dye and a pigment to produce the well-known cobalt blue. There is a view according to which Co was designated by H. Brandt in 1735, and its name stems from German word "*Kobold*." In this connection, I may add that the name appears in a worldwide animation, *Mightly Atom*, created by O. Tezuka, as the name of the brother of the main character, Atom. Moreover, Co has been also widely used as alloys with Fe, Ni, and Cr: Ni-Fe-Co alloy is employed as a binder between glass and metal and Cr-Co-W alloy is dental or surgery material.

The book is motivated by the desire to describe why cobalt oxides have drawn much interest as functional materials, together with their peculiar physical properties partially originating from a rich variety of valences and spin states of Co ions. The leading role of the physical phenomena dealt with in this book is owed to the electron. The electron, discovered by J. J. Thomson in 1897, is a particle that cannot be far resolved under normal conditions and has a wave nature as well. Wave-particle duality was evidenced by experiments using the double split performed by C. Jönsson, P. G. Merli, and A. Tonomura in 1961, 1974, and 1989, respectively.

In the Co oxide system, the strong correlation between electrons generally plays a substantial role, where the conventional one-electron approximation fails. In particular, the characteristics of

Co ions in oxides should be focused on in comparison with other transition metals. This book starts with the basis of one-electron band theory and advances toward the stage of strong electron correlation systems and furthermore progresses to cover up-to-date topics such as huge thermoelectric power, superconductivity, and intrinsic inhomogeneity, etc. This book would be of interest to graduate students and researchers in the fields of physics, chemistry, and materials science. Aside from helping readers in the pencil-and-paper solution of problems, the discussion, which this book aims at developing, may be useful for understanding the essence of functional materials.

Tsuyoshi Takami

Acknowledgments

I acknowledge support from the Foundation of a Grant-in-Aid for Young Scientists (B) (Grant No. 21740251 and No. 24740233), Ando Laboratory, Thermal & Electric Energy Technology Foundation, Japan Society for the Promotion of Science (JSPS) Research Fellowships for Young Scientists, the Sasakawa Scientific Research Grant from the Japan Science Society, and the Research Foundation for the Electrotechnology of Chubu. It's thanks to Emeritus Prof. Uichiro Mizutani, Prof. Hiroshi Ikuta, and Dr. Jun Sugiyama that our achievements were successfully obtained and became worthy to be published as this book. Furthermore, I appreciate Profs. Tsunehiro Takeuchi, Jinguang Cheng, and Masayuki Itoh and Dr. Hiroshi Nozaki for their collaboration. Profs. John B. Goodenough and Jianshi Zhou are also thanked for useful discussions. I am indebted to Mr. Sarabjeet Garcha, senior editorial manager of Pan Stanford Publishing Pte. Ltd., for his advice on form and assistance for preparation. I appreciate Mr. Stanford Chong, director and publisher of Pan Stanford Publishing Pte. Ltd., for inviting me to develop my research topic into this book.

Tsuyoshi Takami

Chapter 1

Introduction

1.1 Prologue

Cobalt oxides are known to exhibit peculiar physical/chemical properties, as well as being suitable candidates for wide applications such as electrode materials, thermoelectric materials, solid oxide fuel cells, and so forth. Such properties are closely related to versatility of the cobalt cations to adopt different valences and spin states in the narrow energy range, which associate with the charge and spin configuration on $3d$ orbitals, respectively. In this chapter, the charge, spin, and orbital for the $3d$ electron system are introduced, and then their role in the transport properties under external perturbations is formulated.

1.2 Charge, Spin, and Orbitals

Since electrons in materials are attracted by an atomic nucleus, their characteristics depend on the kind of elements and orbitals. The elements generally show the tendency given in Table 1.1. The density of conduction electrons $\rho(r)$ at a distance r from a certain atom deviates from the initial $\rho_0(r)$, and the potential $\varphi(r)$ and $\rho(r)$

Functional Cobalt Oxides: Fundamentals, Properties, and Applications
Tsuyoshi Takami
Copyright © 2014 Pan Stanford Publishing Pte. Ltd.
ISBN 978-981-4463-32-4 (Hardcover), 978-981-4463-33-1 (eBook)
www.panstanford.com

Table 1.1 Characteristics of representative elements

	Alkaline metal	Transition metal	Actinoid metal	Rare-earth metal
Orbital	s	$3d$	$5f$	$4f$
Element	K, Na, $\cdots$	Fe, Co, $\cdots$	Pa, U, $\cdots$	Ce, Pr, $\cdots$
Extension of the wave function	large	middle	middle	small
Characteristic	itinerant	itinerant/localized	itinerant	localized
Electron correlation	weak	strong	midrange	strong

are connected with the Poisson equation:

$$\nabla^2 \varphi = -4\pi e^2 \rho(r) - \rho_0(r). \tag{1.1}$$

The proportional relationship of $(2\pi/L)^3 : 2 = 4\pi k_F^3/3 : N_A$ leads to

$$k_F = 3\pi^2 (N_A/V)^{1/3} = (3\pi^2 \rho(r))^{1/3}, \tag{1.2}$$

where L, k_F, N_A, and V are the edge length, the radius of the Fermi sphere, the Avogadro number, and the volume, respectively. Thus,

$$E_F + \varphi(r) = \left(\frac{\hbar}{2m}\right)(3\pi^2 \rho(r))^{2/3} \tag{1.3}$$

holds, which is converted to

$$\rho(r) = \frac{1}{3\pi^2}\left(\frac{2m}{\hbar}\right)^{3/2} E_F^{3/2}\left(1 + \frac{\varphi}{E_F}\right)^{3/2}, \tag{1.4}$$

where E_F and $\hbar$ are the Fermi energy and the Planck's constant, respectively. By applying the Taylor expansion, this equation can be written as

$$\rho(r) = \rho_0\left(1 + \frac{3}{2}\frac{\varphi}{E_F}\right), \tag{1.5}$$

where $\rho_0 = (1/3\pi^2)(2m/\hbar^2)^{3/2} E_F^{3/2}$.

Insertion into Eq. 1.1 gives

$$\nabla^2 \varphi = -\lambda^2 \varphi, \tag{1.6}$$

where $\lambda = (6\pi e^2 \rho_0/E_F)^{1/2}$. This equation is a function of the radius parameter r in the spherical coordinate and is rewritten as

$$\frac{1}{r^2}\frac{d}{dr}\left(r^2 \frac{d\varphi(r)}{dr}\right) = \lambda^2 \varphi(r), \tag{1.7}$$

and thus it can be confirmed that its solution becomes

$$\varphi(r) = -\frac{e^2 \Delta Z \exp(\lambda r)}{r^2}, \tag{1.8}$$

where ΔZ is the difference of the amount of charge. The parameter λ is called the Thomas–Fermi screening parameter. For instance, for the Co element alone, the insertion of $E_F = 7.4$ eV and $\rho_0 = 9.1 \times 10^{22}$ /cm^3 gives $1/\lambda = 0.55$ Å. On the other hand, the intratomic distance equals the lattice constant a of 2.51 Å. Consequently, one can realize how effective the screening effect is because of $1/\lambda \ll a$.

For strongly correlated electron systems, since the scattering effect is less effective, three internal degrees of freedom of electrons, that is, charge, spin, and orbital, can be attributable to physical properties, which cannot be explained by a conventional band picture. When such degeneracies functionate on the crystal lattice, various electronic phases are formed. Charge is one of the properties of an elementary particle, and its quantity takes a positive sign or negative sign corresponding to the electron or the hole, respectively. The spin angular momentum is the angular momentum of elementary particles such as electrons and quarks, and complex particles possess one of the quantum degeneracies. On the other hand, the wave function is called orbital as a quantum dynamical concept corresponding to an orbit for classical electrons. Both the spin and the orbital angular momentum are ascribed to the total angular momentum of particles.

The Schrödinger equation is written as

$$\mathcal{H}\psi(\mathbf{r}) = E\psi(\mathbf{r}), \tag{1.9}$$

using the Hamiltonian defined as

$$\mathcal{H} = -\frac{\hbar^2}{2m_e}\nabla^2 + V(r). \tag{1.10}$$

Now, as long as the wave function is written as $\psi(\mathbf{r}) = f(r)Y(\theta, \varphi)$, the radial and angle wave equations are, respectively, given by

$$\frac{1}{\sin\theta}\frac{\partial}{\partial\theta}\left(\sin\theta\frac{\partial}{\partial\theta}\right) + \frac{1}{\sin^2\theta}\frac{\partial^2}{\partial\varphi^2}Y(\theta, \varphi) + \chi Y(\theta, \varphi) = 0 \tag{1.11}$$

and

$$\frac{1}{r^2}\frac{d}{dr}\left(r^2\frac{d}{dr}\right) + \frac{2m}{\hbar^2}\left(E - V(r) - \frac{\chi}{r^2}\right)f(r) = 0, \tag{1.12}$$

where $\chi \ (= l(l+1))$ is a constant with no dimension. Furthermore, by using the relationship of $Y(\theta, \varphi) = \Theta(\theta)\Phi(\varphi)$, Eq. 1.11 is converted to

$$\frac{d^2\Phi}{d\varphi^2} + v\Phi = 0 \tag{1.13}$$

and

$$\frac{1}{\sin\theta}\frac{d}{d\theta}\left(\sin\theta\frac{d\Phi}{d\theta}\right) + \left(\chi - \frac{v}{\sin^2\theta}\theta\right) = 0. \tag{1.14}$$

The solution of Eq. 1.13 is

$$\Phi = \frac{1}{\sqrt{2\pi}}\exp(im\varphi), \tag{1.15}$$

where $v = m^2$.

On the other hand, Eq. 1.14 is rewritten as

$$\frac{d}{d\omega}\left(1 - \omega^2\right)\frac{dP}{d\omega} + \left(\chi - \frac{m^2}{1 - \omega^2}\right)P = 0, \tag{1.16}$$

where $\omega = \cos\theta$. P is called the associated Legendre function and is specified as

$$P_l^m(\omega) = (1 - \omega^2)^{1/2|m|}\frac{d^{|m|}}{d\omega^{|m|}}P_l(\omega). \tag{1.17}$$

Thus,

$$Y_{l,m}(\theta, \varphi) = C_{l,m}P_l^m(\cos\theta)\Phi(\varphi) \tag{1.18}$$

$$= C_{l,m}P_l^m(\cos\theta)\frac{1}{\sqrt{2\pi}}\exp(im\varphi), \tag{1.19}$$

where

$$C_{l,m} = (-1)^{(m+|m|)/2}\frac{1}{\sqrt{2\pi}}\left[\frac{2l+1}{2}\frac{(l-|m|)!}{(l+|m|)!}\right]^{1/2}. \tag{1.20}$$

The wave function is finally expressed as

$$\varphi_{l,m} = f(r)Y_l^m(\theta, \varphi). \tag{1.21}$$

Therefore, the wave function of $3d$ orbitals for $(l, m) = (2, 0)$ is calculated as

$$\varphi_{2,0}(\mathbf{r}) = (-1)^0 \frac{1}{\sqrt{2\pi}} \sqrt{\frac{5}{2}} f \frac{1}{2^2 2!} (1 - \cos\theta)^0 \frac{d^2}{d^2\cos\theta} (\cos^2\theta - 1)^2 \tag{1.22}$$

$$= \sqrt{\frac{5}{16\pi}} f (3\cos\theta^2 - 1) \tag{1.23}$$

$$= \sqrt{\frac{5}{16\pi}} f \frac{1}{r^2} (3z^2 - r^2). \tag{1.24}$$

Please note here the relationship of $x = r\sin\theta\cos\varphi$, $y = r\sin\theta\sin\varphi$, and $z = r\cos\theta$. A similar procedure brings about

$$\varphi_{2,1\pm}(\mathbf{r}) = \sqrt{\frac{15}{4\pi}} f \frac{1}{r^2} zx, \quad \sqrt{\frac{15}{4\pi}} f \frac{1}{r^2} yz \tag{1.25}$$

and

$$\varphi_{2,2\pm}(\mathbf{r}) = \sqrt{\frac{15}{16\pi}} f \frac{1}{r^2} (x^2 - y^2), \quad \sqrt{\frac{15}{16\pi}} f \frac{1}{r^2} xy. \tag{1.26}$$

Among them, the relative probability fixed by the last wave function, where electrons distribute, is proportional to $(x^2 y^2 =)$ $\sin^4\theta\cos^2\varphi\sin^2\varphi$. As shown in Fig. 1.1a, $\sin^4\theta$ mainly has finite values around $\pi/2$ and $3/2\pi$, which means most electrons disperse on the x-y plane. What is more, the $\cos^2\varphi\sin^2\varphi$ curve in Fig. 1.1b suggests that electrons spread over toward a diagonal-line direction (see also $3d_{xy}$ in Fig. 1.2). Five orbitals $\varphi_{2,0}(\mathbf{r})$, $\varphi_{2,1\pm}(\mathbf{r})$, and $\varphi_{2,2\pm}(\mathbf{r})$ are degenerate, but they are known to split into lower levels (t_{2g}) and higher levels (e_g) in the case of the ideal MO_6 octahedron due to the crystal field splitting. This orbital splitting can be interpreted intuitively as follows: $3z^2$-r^2 and x^2-y^2 orbitals in the e_g level turn to the position of O^{2-} ions (see Fig. 1.2). Coulomb interaction becomes strong due to negative charges at a close range, resulting in the e_g orbitals relatively going up compared to the t_{2g} orbitals.

More quantitatively, in the case of trigonal symmetry, the whole electric field is

$$V(\mathbf{r}) = V_2^0 + V_4^0 + V_4^3 \tag{1.27}$$
$$= A_2^0(3z^2 - r^2) + A_4^0(35z^4 - 30z^2r^2 + 3r^4) + A_4^3(x^3 - 3xy^2)z \tag{1.28}$$

and each component is written as

$$\sum (3z^2 - r^2) = \alpha < r^2 > [3L_z^2 - L(L+1)], \tag{1.29}$$

$$\sum (35z^4 - 30z^2r^2 + 3r^4) = \beta < r^4 > [35L_z^4 - 30L(L+1)L_z^2$$
$$+ 25L_z^2 - 6L(L+1) + 3L^2(L+1)^2], \tag{1.30}$$

and

$$\sum (x^3 - 3xy^2)z = \frac{\beta}{4} < r^4 > [L_z(L_+^3 + L_-^3) + (L_+^3 + L_-^3)L_z]. \tag{1.31}$$

Hereafter, let's calculate the detailed energy level. First we estimate V_4^3, including the matrix element.

$$L_+^2 = \begin{pmatrix} 0 & 2 & 0 & 0 & 0 \\ 0 & 0 & \sqrt{6} & 0 & 0 \\ 0 & 0 & 0 & \sqrt{6} & 0 \\ 0 & 0 & 0 & 0 & 2 \\ 0 & 0 & 0 & 0 & 0 \end{pmatrix} \begin{pmatrix} 0 & 2 & 0 & 0 & 0 \\ 0 & 0 & \sqrt{6} & 0 & 0 \\ 0 & 0 & 0 & \sqrt{6} & 0 \\ 0 & 0 & 0 & 0 & 2 \\ 0 & 0 & 0 & 0 & 0 \end{pmatrix} = \begin{pmatrix} 0 & 0 & 2\sqrt{6} & 0 & 0 \\ 0 & 0 & 0 & 6 & 0 \\ 0 & 0 & 0 & 0 & 2\sqrt{6} \\ 0 & 0 & 0 & 0 & 0 \\ 0 & 0 & 0 & 0 & 0 \end{pmatrix} \tag{1.32}$$

$$L_+^3 = \begin{pmatrix} 0 & 0 & 0 & 12 & 0 \\ 0 & 0 & 0 & 0 & 12 \\ 0 & 0 & 0 & 0 & 0 \\ 0 & 0 & 0 & 0 & 0 \\ 0 & 0 & 0 & 0 & 0 \end{pmatrix} \tag{1.33}$$

A symmetric procedure on the matrix elements brings about

$$L_-^3 = \begin{pmatrix} 0 & 0 & 0 & 0 & 0 \\ 0 & 0 & 0 & 0 & 0 \\ 0 & 0 & 0 & 0 & 0 \\ 12 & 0 & 0 & 0 & 0 \\ 0 & 12 & 0 & 0 & 0 \end{pmatrix}. \tag{1.34}$$

The first and second terms in Eq. 1.31, respectively, become

$$L_z(L_+^3 + L_-^3) = \begin{pmatrix} 2 & 0 & 0 & 0 & 0 \\ 0 & 1 & 0 & 0 & 0 \\ 0 & 0 & 0 & 0 & 0 \\ 0 & 0 & 0 & -1 & 0 \\ 0 & 0 & 0 & 0 & -2 \end{pmatrix} \begin{pmatrix} 0 & 0 & 0 & 12 & 0 \\ 0 & 0 & 0 & 0 & 12 \\ 0 & 0 & 0 & 0 & 0 \\ 12 & 0 & 0 & 0 & 0 \\ 0 & 12 & 0 & 0 & 0 \end{pmatrix}$$

$$= \begin{pmatrix} 0 & 0 & 0 & 24 & 0 \\ 0 & 0 & 0 & 0 & 12 \\ 0 & 0 & 0 & 0 & 0 \\ -12 & 0 & 0 & 0 & 0 \\ 0 & -24 & 0 & 0 & 0 \end{pmatrix} \tag{1.35}$$

and

$$(L_+^3 + L_-^3)L_z = \begin{pmatrix} 0 & 0 & 0 & -12 & 0 \\ 0 & 0 & 0 & 0 & -24 \\ 0 & 0 & 0 & 0 & 0 \\ 24 & 0 & 0 & 0 & 0 \\ 0 & 12 & 0 & 0 & 0 \end{pmatrix}. \tag{1.36}$$

Thus,

$$L_z(L_+^3 + L_-^3) + (L_+^3 + L_-^3)L_z = \begin{pmatrix} 0 & 0 & 0 & 12 & 0 \\ 0 & 0 & 0 & 0 & -12 \\ 0 & 0 & 0 & 0 & 0 \\ 12 & 0 & 0 & 0 & 0 \\ 0 & -12 & 0 & 0 & 0 \end{pmatrix}. \tag{1.37}$$

For easy calculation of the eigenvalue, the matrix elements are rearranged in order of $M = 0, 1, -2, -1,$ and 2 instead of the present alignment of $M = 2, 1, 0, -1,$ and -2.

$$L_z(L_+^3 + L_-^3) + (L_+^3 + L_-^3)L_z = \begin{pmatrix} 0 & 0 & 0 & 0 & 0 \\ 0 & 0 & -12 & 0 & 0 \\ 0 & -12 & 0 & 0 & 0 \\ 0 & 0 & 0 & 0 & 12 \\ 0 & 0 & 0 & 12 & 0 \end{pmatrix}. \tag{1.38}$$

Here we replace $-3A_2^0\alpha < r^2 >= A$, $12A_4^0\beta < r^4 >= B$, and $-3A_4^3\beta < r^4 >= C$. The two residual electric fields are expressed as follows:

$$V_2^0 = \alpha < r^2 > A_2^0(3M^2 - 6) \tag{1.39}$$

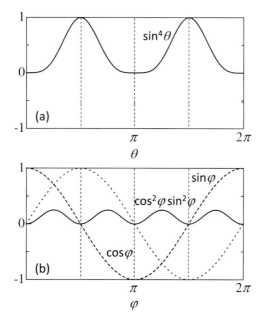

Figure 1.1 Functions of (a) $\sin^4\theta$ and (b) $\cos^2\varphi\,\sin^2\varphi$ together with $\sin\varphi$ (dotted curve) and $\cos\varphi$ (dashed curve).

and

$$V_4^0 = \beta < r^4 > A_4^0(35M^4 - 155M^2 + 72), \qquad (1.40)$$

and thereby

$$V_2^0 = \begin{cases} 2A & (M = 0) \\ A & (M = 1) \\ -2A & (M = -2) \\ A & (M = -1) \\ -2A & (M = 2) \end{cases} \qquad (1.41)$$

and

$$V_4^0 = \begin{cases} 6B & (M = 0) \\ -4B & (M = 1) \\ B & (M = -2) \\ -4B & (M = -1) \\ B & (M = 2). \end{cases} \qquad (1.42)$$

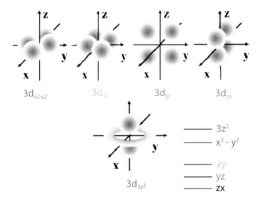

Figure 1.2 Schematic illustration of $3d$ orbitals when the transition metal (M) is put in the ideal MO_6 octahedron.

Finally, the Hamiltonian in a crystal field is expressed as

$$\mathcal{H} = \begin{pmatrix} 2A + 6B & 0 & 0 & 0 & 0 \\ 0 & A - 4B & C & 0 & 0 \\ 0 & C & -2A + B & 0 & 0 \\ 0 & 0 & 0 & A - 4B & -C \\ 0 & 0 & 0 & -C & -2A + B \end{pmatrix}. \quad (1.43)$$

In the case of a cubic electric field, the relationships $A = 0$ and $C = 5\sqrt{2}A_4^0$ are held. The eigenvalue derived from $\mathcal{H}\varphi = E\varphi$ is

$$E_0 = 6B \quad \text{(three-hold degeneracy)} \quad (1.44)$$

and

$$E_1 = -9B \quad \text{(two-hold degeneracy)}. \quad (1.45)$$

The energy levels expressed by Eqs. 1.44 and 1.45 correspond to t_{2g} and e_g orbitals, respectively.

Let us next attempt to extend orbital levels into spin configurations. Electrons with the spin angular momentum are distributed, as shown in Fig. 1.3. For example, the Co atom possesses the electron configuration of $(1s)^2(2s)^2(2p)^6(3s)^2(3p)^6(3d)^7(4s)^2$, in which the characters in ( ) and the upper right of ( ) denote the orbital and the number of electrons, respectively. As can be speculated from the name "transition" electrons are delivered to $4s$ orbitals prior to $3d$ orbitals for the 19th element K and above. When Co becomes

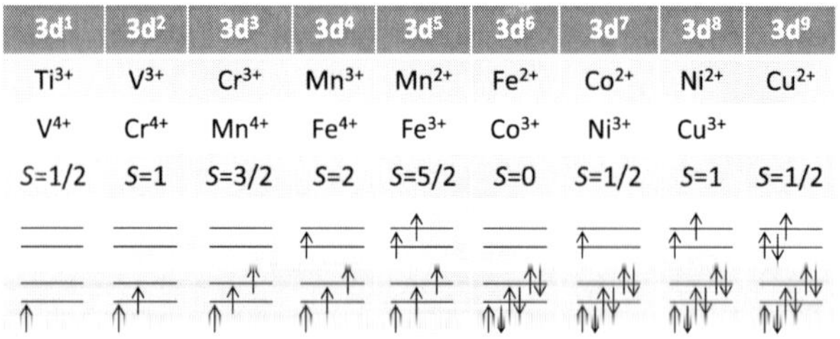

3d¹	3d²	3d³	3d⁴	3d⁵	3d⁶	3d⁷	3d⁸	3d⁹
Ti^{3+}	V^{3+}	Cr^{3+}	Mn^{3+}	Mn^{2+}	Fe^{2+}	Co^{2+}	Ni^{2+}	Cu^{2+}
V^{4+}	Cr^{4+}	Mn^{4+}	Fe^{4+}	Fe^{3+}	Co^{3+}	Ni^{3+}	Cu^{3+}	
$S=1/2$	$S=1$	$S=3/2$	$S=2$	$S=5/2$	$S=0$	$S=1/2$	$S=1$	$S=1/2$

Figure 1.3 $3d$ electron configuration of the transition metal (M) ions in the MO_6 octahedron. There are several ions with multivalences, such as Mn, Fe, and Co ions. S and the arrow represent the spin angular momentum and the spin, respectively.

trivalent, three electrons are directly reduced from the $4s$ and $3d$ orbitals. As a result, the number of $3d$ electrons becomes 6, denoted as $(3d)^6$. The situation where odd electrons occupy the e_g orbitals could lead to further splitting of the e_g orbitals to stabilize the energy of the whole system. Such an electron–phonon effect to lower the symmetry of the crystal field is called the Jahn–Teller effect [7]. The spin state for such spin configurations could be specified to some extent by investigating the presence or absence of the crystal distortion.

Next, let's extend the electronic state of Co itself to that of the CoO_6 octahedral cluster. Co supplies $3d$, $4s$, and $4p$ orbitals as an outer orbital, while O does the $2p$ orbital. The O $2p$ orbital in the O_6^{-12} cluster is composed of seven irreducible expressions, that is, $a_{1g}, t_{1u}, t_{1u'}, t_{2u}, t_{2g}, t_{1g},$ and e_g, and they possess 1-, 3-, 3-, 3-, 3-, 3-, and 2-hold degeneracy, respectively. To form the molecular orbital by hybridization of two atomic orbitals, (i) two energy levels of atomic orbitals, (ii) well overlapping of two atomic orbitals, and (iii) the same symmetry for two atomic orbitals toward the molecular axis are required. As can be seen from Fig. 1.4, the molecular orbitals consists of bonding, antibonding, and nonbonding orbitals, and antibonding orbitals t_{2g}^* and e_g^* are found to dominate the electronic properties. If I may add a few words, the chief ingredient of the bonding orbitals a_{1g} and t_{1u} is oxygen. Thus, this covalent bonding is considered to take part in stabilizing energetically the perovskite structure.

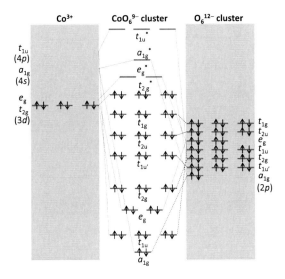

Figure 1.4 Molecular orbital energy level for the CoO_6 octahedral cluster.

So far, we have found that the transition metal $3d$ and the oxygen $2p$ orbitals play an important role. Various electronic structures are realized by a balance among the band width W, the charge transfer energy Δ to transfer electrons between the $2p$ band and the $3d$ band, and the Coulomb energy U, which are shown in order of relative position of the oxygen $2p$ band in Fig. 1.5. Left and right sides are the partial density of states (DOS) of the oxygen $2p$ band and the transition metal $3d$ band, respectively. Here, the $3d$ band in the right splits into a lower Hubbard band and an upper Hubbard band due to Coulomb interaction. For Fig. 1.5a and Fig. 1.5e, the metallic state is formed. For Fig. 1.5b and Fig. 1.5d, on the other hand, the charge transfer insulating and Mott insulating states are actualized in the case of $\Delta < U$ and $\Delta > U$, respectively.

A well-known Zaanen–Sawatzky–Allen diagram is shown in Fig. 1.6. Figure 1.5 corresponds one to one to this figure [5]. The Mott–Hubbard insulator undergoes a continuous transition to the $3d$ band metal for $U < W_d$ irrespective of the value of Δ. When $\Delta < (W_p + W_d)/2$, the bands coming from $2p$ and $3d$ orbitals start to overlap, resulting in a metallic state.

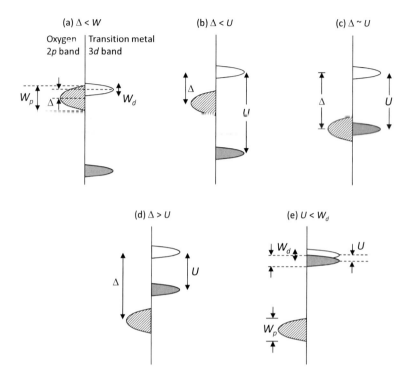

Figure 1.5 Various electronic structures corresponding to the magnitude of W, U, and Δ, where they represent the band width, Coulomb energy, and charge transfer energy, respectively.

1.3 Physical Properties under External Perturbations

Under the electrical field $\mathbf{E}$ and temperature gradient ∇T as external perturbations, the electrical current density $\mathbf{J}$ and thermal current density $\mathbf{U}$ are expressed as

$$\mathbf{J} = A\mathbf{E} + B\nabla T \tag{1.46}$$

and

$$\mathbf{U} = C\mathbf{E} + D\nabla T, \tag{1.47}$$

respectively, using the coefficients A-D.

Let's consider the temperature gradient ∇T between two ends of the sample, but the electric circuit is open so the relationship $\mathbf{J} = 0$ holds. Then, Eq. 1.46 leads to

$$\mathbf{E} = -B/A\nabla T. \tag{1.48}$$

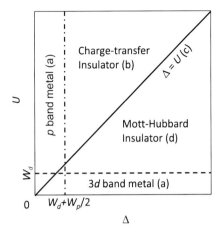

Figure 1.6   Zaanen–Sawatzky–Allen diagram. Panels (a–e) correspond one to one to the electronic states in Fig. 1.5a–e [5].

The coefficient $-(B/A)$ is called the absolute thermoelectric power. When Eqs. 1.46 and 1.47 are adopted under the same condition, the relationship

$$\mathbf{U} = -(C\,B/A - D)\nabla T \qquad (1.49)$$

is induced. The coefficient $C\,B/A - D$ corresponds to the thermal conductivity.

For the moment let us look closely at the thermoelectric power by the physical quantity using the Boltzmann transport equation [8]. The Boltzmann equation under the condition that no magnetic field is applied can be written as

$$-\left(\frac{\partial f}{\partial t}\right)_{\text{scatter}} = \left(-\frac{\partial f_0}{\partial \epsilon}\right)\mathbf{v_k}\cdot\left\{-\left(\frac{\epsilon(\mathbf{k})-\zeta}{T}\right)\nabla T\right.$$
$$\left.+(-e)\left(\mathbf{E}-\frac{\nabla\zeta}{(-e)}\right)\right\}. \qquad (1.50)$$

Here f, f_0, $\mathbf{v_k}$, and ζ represent the electron distribution, the Fermi–Dirac function, the velocity of the electron, and the chemical potential, respectively. First of all, we sort the nature of the Fermi–Dirac function, which determines the distribution of electronic states or the Fermi surface at finite temperatures. The Fermi–Dirac function is expressed in the form of

$$f = \frac{1}{1 + \exp(E - E_{\text{F}}/k_{\text{B}}T)}, \qquad (1.51)$$

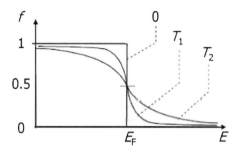

Figure 1.7 Fermi–Dirac distribution function at $T = 0$, T_1, and T_2 $(0 < T_1 < T_2)$ assuming no change in the Fermi energy E_F with temperature.

where E_F is the Fermi energy. Its slope variation with temperature is described in Fig. 1.7. f becomes a step function at approximately absolute zero temperature: $f = 1$ $(E \leq E_F)$ and $f = 0$ $(E > E_F)$. Put it another way, all seats are filled by electrons for $E \leq E_F$, while they are empty for $E > E_F$. With increasing temperature, Eq. 1.51 yields $0 < f < 1$ over $k_B T$ in the vicinity of E_F, indicative of the coexistence of occupied and unoccupied states. With further increase of temperature, the smeared region is found to expand more and more holding $f = 1/2$ at E_F. Please keep in mind that E_F of typical metal is of the order of several eV, which is far larger than $k_B T$, say, merely 25 meV even at room temperature. The change of the electron distribution due to scattering is generally written as

$$\left(\frac{\partial f}{\partial t}\right)_{\text{scatter}} = \int \left\{ f(\mathbf{k}')(1 - f(\mathbf{k})) - f(\mathbf{k})(1 - f(\mathbf{k}')) \right\} Q(\mathbf{k}, \mathbf{k}') d\mathbf{k}',$$

$$(1.52)$$

where $Q(\mathbf{k}, \mathbf{k}')$ denotes the transition probability in the scattering process. We employ the relaxation time approximation for avoiding a formidable calculation. The scattering term is then expressed as

$$-\left(\frac{\partial f}{\partial t}\right)_{\text{scatter}} = \frac{f(\mathbf{r}, \mathbf{k}) - f_0(\epsilon_k, T)}{\tau}, \qquad (1.53)$$

where τ is the relaxation time.

Equation 1.52 is rewritten by inserting Eq. 1.53:

$$f(\mathbf{r}, \mathbf{k}) - f_0(\epsilon_k, T) = \left(-\frac{\partial f_0}{\partial \epsilon}\right) \tau \mathbf{v}_k \cdot \left\{ -\left(\frac{\epsilon(\mathbf{k}) - \zeta}{T}\right) \nabla T \right.$$

$$\left. + (-e) \left(\mathbf{E} - \frac{\nabla \zeta}{(-e)} \right) \right\}. \qquad (1.54)$$

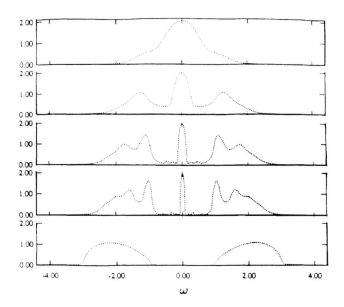

Figure 1.8 Coulomb interaction variation of the DOS; the interaction becomes larger downward [5].

On the other hand, since **J** is written as

$$J = \frac{(-e)}{4\pi^3} \int v_k \{f(r, k) - f_0(\epsilon_k, T)\} \, dk. \tag{1.55}$$

By inserting Eq. 1.54, we further obtain **J** as

$$J = \frac{(-e)}{4\pi^3} \int v_k \left(-\frac{\partial f_0}{\partial \epsilon}\right) \tau v_k \cdot \left\{-\left(\frac{\epsilon(k) - \zeta}{T}\right) \nabla T \right.$$
$$\left. + (-e)\left(E - \frac{\nabla \zeta}{(-e)}\right)\right\} dk. \tag{1.56}$$

Here the following relationship

$$K_n = -\frac{1}{3} \int v_k^2 \tau(\epsilon_k)(\epsilon_k - \zeta)^n \left(\frac{\partial f_0}{\partial \epsilon}\right) dk$$
$$= -\int (\epsilon_k - \zeta)^n \sigma(\epsilon) \left(\frac{\partial f_0}{\partial \epsilon}\right) d\epsilon \tag{1.57}$$

yields the simple formulae of the electrical conductivity and the thermoelectric power:

$$\sigma = \int \sigma(\epsilon) \left(-\frac{\partial f}{\partial \epsilon}\right) d\epsilon = K_0 \tag{1.58}$$

and

$$S = \frac{1}{eT} \frac{\int (\epsilon - \zeta)\sigma(\epsilon)\frac{\partial f_0}{\partial \epsilon} d\epsilon}{\int \sigma(\epsilon)\frac{\partial f_0}{\partial \epsilon} d\epsilon} = \frac{1}{eT} \frac{K_1}{K_0}, \tag{1.59}$$

respectively.

The spectral conductivity $\sigma(\epsilon)$ is expressed as

$$\sigma(\epsilon) = e^2 \sum_{\mathbf{k}} v_G^2(\mathbf{k})\tau(\mathbf{k})\delta(\epsilon(\mathbf{k}) - \epsilon) \tag{1.60}$$

where v_G and δ are the group velocity and the Kronecker delta, respectively. The anisotropies of the crystal structure and the electron correlation are reflected in $v_G(\mathbf{k})$ and $\tau(\mathbf{k})$, respectively, inclusive of temperature variation. Typical patterns of the $\rho(T)$ curve are as follows:

(i) electron–impurity scattering $\cdots$ $\rho \propto T^0$,
(ii) electron–electron scattering $\cdots$ $\rho \propto T^2$, and
(iii) electron–phonon scattering $\cdots$ $\rho \propto T^5$ $(T/\Theta_D < 0.1)$
    $\cdots$ $\rho \propto T$ $(T/\Theta_D > 0.2)$,

where Θ_D is the Debye temperature.

Because Eq. 1.59 includes the Fermi–Dirac distribution function, the formula

$$\int f(E, T)\frac{dF(E)}{dE} dE = F(E_F(T)) + \frac{\pi^2}{6}(k_B T)^2 \left(\frac{d^2 F(E)}{dE^2}\right)_{E=E_F(T)} + \cdots \tag{1.61}$$

can be employed for this part, and then the numerator is reduced to the expansion formula:

$$(\epsilon - \zeta)\sigma(\epsilon) + \frac{\pi^2}{6}(k_B T)^2 \left\{\left((\epsilon - \zeta)\left(\frac{\partial^2 \sigma(\epsilon)}{\partial \epsilon^2}\right)\right)_{E=E_F(T)} + 2\frac{\partial \sigma(\epsilon)}{\partial \epsilon}\right\}$$

$$= -\frac{\pi^2}{3}(k_B T)^2 \frac{\partial \sigma(\epsilon)}{\partial \epsilon}. \tag{1.62}$$

Please note here that the left term of Eq. 1.61 is expanded to

$$[f(\epsilon)F(\epsilon)]_0^\infty - \int_0^\infty F(\epsilon)\left(\frac{df(\epsilon)}{d\epsilon}\right) d\epsilon = \int_0^\infty F(\epsilon)\left(\frac{df(\epsilon)}{d\epsilon}\right) d\epsilon \tag{1.63}$$

and Eq. 1.59 is also applicable to the derivative of the Fermi–Dirac function, as in the case of this procedure. Therefore, the thermoelectric power is finally formulated as

$$S = \frac{\pi^2}{3(-e)} k_B^2 T \left(\frac{\partial \ln \sigma(\epsilon)}{\partial \epsilon} \right). \tag{1.64}$$

Here, $\sigma(\epsilon)$ is proportional to the DOS; Eq. 1.64 implies a small DOS with a steep slope causes a large S. So, the variation of DOS with the Coulomb interaction U attracts one's attention. We see, hinted in Fig. 1.8, how U affects DOS [5]. For a small U, the DOS is smooth and continuous. However, the DOS starts to split with increasing U, and finally the gap is formed near $\omega = 0$ (Mott insulator). It is emphasized here that the DOS with a significant sharp slope appears just before opening the gap.

Let us how look at the thermoelectric power in a different angle. The constant C in Eq. 1.47 can be rewritten as σST using σ, S, and T assuming quite a small temperature gradient, which gives the relationship

$$\frac{\mathbf{U}}{T} = S\mathbf{J}. \tag{1.65}$$

The left-hand term means the entropy current density, and thus the thermoelectric power is regarded as the ratio of the entropy current to the electrical current. Given the same conduction mechanism both for $\mathbf{U}$ and $\mathbf{J}$, the thermoelectric power is found to become an entropy per charge carrier.

This viewpoint can be also simply discussed on the basis of the thermodynamics. The free energy of the sample must be same everywhere in the steady state. The sum of the kinetic energy and potential energy is same for two ends of the sample, and thus the equation

$$\mu(T_c) + eV(T_c) = \mu(T_h) + eV(T_h) \tag{1.66}$$

holds, where V, T_h, and T_c show the potential, the temperature of the high-temperature side, and the temperature of the low-temperature side, respectively. From the definition of thermoelectric power, its expression is calculated in the limit of a vanishing temperature gradient:

$$S = \frac{V(T_h) - V(T_c)}{T_h - T_c} = -\frac{1}{e} \frac{\partial \mu}{\partial T} = -\frac{1}{e} \frac{s}{N}. \tag{1.67}$$

Here, s and N represent the entropy and the number of electrons, respectively. As $T \to \infty$, the entropy is written as

$$s = k_B \ln g, \tag{1.68}$$

where g is the number of possible states. Hence, the thermoelectric power in the high-temperature limit is given by

$$S = -\frac{k_B}{e} \frac{\partial \ln g}{\partial N}, \tag{1.69}$$

which is expanded to the generalized Heikes formula [12, 13], that is, for U, $T \ll k_B T$, g is given by

$$g = {}_{2N_A}C_N = \frac{(2N_A!)}{N!(2N_A - N)!}, \tag{1.70}$$

where N_A is the system size. Expansion by means of the Stirlings approximation leads to

$$\ln g = \ln(2N_A)! - \ln N! - \ln(2N_A - N)! \tag{1.71}$$

$$= 2N_A\ln(2N_A) - 2N_A - (N\ln N - N) - (2N_A - N)\ln(2N_A - N) \tag{1.72}$$

$$- (2N_A - N)$$

and its partial differentiation becomes

$$\frac{\partial g}{\partial N} = -(\ln N + N\frac{1}{N} - 1) - -\ln(2N_A - N) - (2N_A - N)\frac{1}{2N_A - N} \tag{1.73}$$

$$= \frac{2 - N/N_A}{N/N_A}. \tag{1.74}$$

Thus, the thermoelectric power becomes

$$S = -\frac{k_B}{e} \ln\left(\frac{2 - \rho}{\rho}\right), \tag{1.75}$$

with $\rho = N/N_A$ using the system size as a denominator. Let's consider the situation where the system is composed of two different sites, for example, Co^{3+} and Co^{4+}. The degeneracy of both ions and the number of ways of arranging these states yield

$$g = g_3^{N_A - M} g_4^M \frac{N_A!}{M!(N_A - M)!}, \tag{1.76}$$

where g_3, g_4, and M are the number of the configurations of Co^{3+}, the number of the configurations of Co^{4+}, and the number of Co^{4+}, respectively. The thermoelectric power is eventually expressed by

$$S = -\frac{k_B}{e} \ln\left(\frac{g_3}{g_4} \frac{x}{1 - x}\right). \tag{1.77}$$

x $(= M/N_A)$ in the equation is the concentration of Co^{4+}, in other words, the amount of introduced carriers when $g_3 = 1$ (LS Co^{3+}). I might incidentally remark that $k_B/e \approx 86.17$ $\mu V/K$ and S can be controlled by g_3, g_4, and x in the logarithm term. This is the extended Heikes formula advanced by Koshibae *et al.* [8, 9].

Next, let's formulate the electronic thermal conductivity by using the Boltzmann transport equation, granted the relaxation time approximation. The thermal current density $\mathbf{U}$ is expressed as

$$\mathbf{U} = \frac{1}{4\pi^3} \int \mathbf{v_k}\epsilon(\mathbf{k}) f(\mathbf{r}, \mathbf{k}) d\mathbf{k} \left\{ \frac{\mathbf{J}}{(-e)} \right\} \zeta. \tag{1.78}$$

Since the relationships $\int \mathbf{v_k} f_0(\mathbf{k}) d\mathbf{k} = 0$ and $\int \epsilon_{\mathbf{k}} \mathbf{v_k} f_0(\mathbf{k}) d\mathbf{k} = 0$ are held, the above equation can be transformed to

$$\mathbf{U} = \frac{1}{4\pi^3} \int [\epsilon(\mathbf{k}) - \zeta] \mathbf{v_k} [f(\mathbf{r}, \mathbf{k}) - f_0(\mathbf{k})] d\mathbf{k}. \tag{1.79}$$

By inserting Eq. 1.54 into Eq. 1.79, the thermal conductivity is expressed as

$$\mathbf{U} = \frac{1}{4\pi^3} \int [\epsilon(\mathbf{k}) - \zeta]^2 \tau(\mathbf{k}) \mathbf{v_k} \mathbf{v_k} \left(-\frac{\partial f_0}{\partial \epsilon} \right) d\mathbf{k} \frac{\nabla T}{T}. \tag{1.80}$$

Since the above equation includes the derivative of the Fermi–Dirac distribution function, Eq. 1.61 is available, and then we obtain

$$\mathbf{U} = - \left[\frac{(\epsilon(\mathbf{k}) - \zeta)^2 \sigma(\epsilon)}{(-e)^2} + \frac{\pi^2 (k_B T)^2}{6} \frac{\partial^2}{\partial \epsilon^2} \left((\epsilon(\mathbf{k}) - \zeta)^2 \frac{\sigma(\epsilon)}{(-e)^2} \right) \right. $$
$$\left. + \cdots \right] \left(\frac{\nabla T}{T} \right). \tag{1.81}$$

The insertion of $\epsilon = \zeta$ leads to the relationship

$$\mathbf{U} = -\frac{\pi^2 k_B^2 T \sigma}{3(-e)^2} \nabla T. \tag{1.82}$$

As already shown, the thermal conductivity is $-D + BC/A$ $(= -D(1 - BC/AD))$. Here, the second term is of the order of $(T/T_F)^2$, and its value is negligibly small. Therefore, the electronic thermal conductivity corresponding to the coefficient of Eq. 1.82 is finally expressed as

$$\kappa_{el} = \frac{\pi^2 k_B^2 T}{3(-e)^2} \sigma(T). \tag{1.83}$$

This equation means that the ratio of the electronic thermal conductivity to the electrical resistivity becomes constant, which is referred to as the Wiedemann–Franz law. In particular, we need to reconfirm the universality of the above relationship. Two conditions are needed to be fulfilled, that is, (i) the elastic or quasi-elastic scattering of electrons and (ii) the energy independent relaxation time. Namely keep in mind that the Wiedemann–Franz law is actually valid below temperatures where the residual resistivity is observed and above the Debye temperature Θ_D where the thermal energy $k_B T$ smears the maximum phonon energy $k_B \Theta_D$.

Historically, Wiedemann and Franz pointed out that the ratio of the thermal conductivity and the electrical resistivity around room temperature is constant for many pure metals in 1853 [10]. Drude reported the theoretical evidence for application of the Wiedemann–Franz law; it was before the birth of quantum mechanics [11]. According to the kinetic theory of gases, thermal conductivity is given by

$$\kappa = \frac{1}{3}\lambda v c_v, \tag{1.84}$$

where λ, v, and c_v are the mean free path, the velocity, and the specific heat at constant volume, respectively. Among three factors, the velocity is ordinarily considered to be almost constant. Hereafter let's formulate c_v as the phonon specific heat. The internal energy of the phonon system is expressed as the integral of the product of the energy $\hbar\omega$, the DOS $D(\omega)$, and the Planck distribution function $n(\omega)$:

$$U_{\text{lattice}} = \int_0^{\omega_D} \hbar\omega D(\omega)n(\omega)d\omega. \tag{1.85}$$

The number of lattice wave in the sphere with a radius of q in reciprocal space is

$$\frac{4\pi q^3}{3} : N(q) = \left(\frac{2\pi}{L}\right)^3 : 3, \tag{1.86}$$

namely,

$$N(q) = \frac{V}{2\pi^2}q^3. \tag{1.87}$$

By using Eq. 1.85 and the assumption under the Debye model, $\omega = sq$, the DOS of phonons becomes

$$D(\omega)d\omega = \left(\frac{dN(q)}{dq}\right)\left(\frac{dq}{d\omega}\right)d\omega \qquad (1.88)$$

$$= \left(\frac{3q^2}{2\pi^2}\right)\left(\frac{1}{s}\right)d\omega \qquad (1.89)$$

$$= \left(\frac{3\omega^2}{2\pi^2}\right)\left(\frac{1}{s^3}\right)d\omega. \qquad (1.90)$$

An insertion of Eq. 1.90 and the parameter conversion of $x = \hbar\omega/k_B T$ makes the internal energy as follows:

$$U_{\text{lattice}} = \frac{3k_B^4 T^4}{2\pi^2\hbar^3 s^3}\int_0^{x_D}\frac{x^3}{e^x - 1}dx. \qquad (1.91)$$

Thus, the phonon specific heat leads to

$$C_{\text{lattice}} = \frac{\partial U_{\text{lattice}}}{\partial T} \qquad (1.92)$$

$$= \frac{3k_B^4}{2\pi^2\hbar^3 s^3}\left(4T^3\int_0^{x_D}\frac{x^3}{e^x - 1}dx - T^3\frac{x_D^4}{e^{x_D} - 1}\right). \qquad (1.93)$$

Here the relationship of integration by parts $\int_0^x y^4 e^y/\ (e^y - 1)dy = \int_0^x -1/(e^y - 1)'y^4 dy = -x^4/(e^x - 1) + 4\int_0^x y^3/\ (e^y - 1)dy$ is applicable. Finally, the phonon specific heat is found to be

$$C_{\text{lattice}} = 9N_A k_B \left(\frac{T}{\Theta_D}\right)^3\int_0^{x_D}\frac{e^x x^4}{(e^x - 1)^2}dx, \qquad (1.94)$$

in which the relationships of $N_A = N/V$ and $\Theta_D = (\frac{\hbar s}{k_B})(6\pi^2\frac{N}{V})^{1/3}$ are applied in the derivation process. For $T \ll \Theta_D$, the upper limit of the integral is considered infinite and the integral becomes constant. Therefore, one can confirm C_{lattice} is proportional to T^3.

λ of electrons at lower temperatures (≤ 10 K) is determined by the impurity concentration, and thus shows no temperature dependence. As is well known, the electronic specific heat is proportional to temperature. As a result, the electron thermal conductivity κ_{el} is proportional to temperature. With increasing temperature, the number of phonons increases in proportion as T^3 and as T at $T \leq \Theta_D$ and $T \geq \Theta_D$, respectively. λ is proportional to the inverse of the number of phonons, resulting in the T^{-2} and T^0 variation of κ_{el} (see Fig. 1.9a).

Here let's digest briefly the temperature variation of the number of phonons. The number of phonons per unit volume can be estimated from

$$n = \int_0^{\omega_D} D(\omega)n(\omega, T)d\omega, \tag{1.95}$$

where $D(\omega)$ and $n(\omega, T)$ are the phonon DOS and the Planck distribution function. Given the Debye model, $D(\omega) \propto \omega^2$, and thus

$$n \propto \int_0^{\omega_D} \frac{\omega^2}{e^{\hbar\omega/k_B T}} d\omega. \tag{1.96}$$

Replacement of $\hbar\omega/k_B T$ with x immediately leads to

$$n \propto (k_B T)^3 \int_0^{x_D} \frac{x^2}{e^x - 1} dx. \tag{1.97}$$

For $T < \Theta_D$, the integral part becomes a definite integral because of the far larger value of the upper limit, x_D. Ultimately, n increases with temperature in proportion as T^3. For $T > \Theta_D$, on the other hand, we expand the exponential part:

$$n \propto (k_B T)^3 \int_0^{x_D} \frac{x^2}{(1 + x + \cdots) - 1} dx \tag{1.98}$$

$$\propto (k_B T)^3 \int_0^{x_D} x\, dx \tag{1.99}$$

$$\propto k_B T; \tag{1.100}$$

that is, n behaves showing a T linear dependence.

For the phonon thermal conductivity (κ_{ph}), λ is constant at lower temperatures due to impurity and lattice vacancy and shows T^{-1} dependence at higher temperatures by phonon–phonon scattering. The phonon specific heat obeys T^3 at lower temperatures and approaches a constant value at high temperature, as specified in the Dulong–Petit law. κ_{ph}, therefore, increases in proportion as T^3 at lower temperatures and decreases as T^{-1} at higher temperatures, as shown in Fig. 1.9b.

I will shift the topic away from electronic and phonon thermal conductivity to spin thermal conductivity in light of its temperature dependence. The mechanism of spin thermal conductivity κ_{spin} is not well established, but its temperature variation can be expected to predict basically from the λ and c_v behavior as well, especially

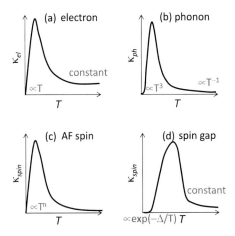

Figure 1.9 Temperature dependences of thermal conductivity of (a) electron, (b) phonon, (c) spin in an n-dimensional system, and (d) spin in a spin gap system.

$c_V(T)$ at lower temperature. For example, the magnetic specific heat is proportional to T^n, where n is the number of spin networks formed by connecting the parts with strong interspin interaction, for example, κ_{spin} is linear to temperature for $n = 1$. On the other hand, for a spin gap system where there is a spin gap in the excitation energy of the spin system, κ_{spin} is proportional to $\exp(-\Delta/T)$, where Δ is the gap energy (see Fig. 1.9d).

Historically, the report that the thermal conductivity by spins is determined by thermal diffusion exists in 1958 [12]. Namely, spin excitation is limited in the interspin distance and κ_{spin} is pretty small. However, Huber *et al.* pointed out that λ of spin excitation becomes infinite, which has brought about active research on the spin thermal conductivity again [13]. From an experimental aspect, κ_{spin} has been discovered one after another in materials such as $Sr_{14}Cu_{24}O_{41}$, Sr_2CuO_4, and La_2CuO_4 [14, 15]. The conditions where κ_{spin} is observable are: (i) large intraspin network exchange interaction J; (ii) small interspin network exchange interaction J'; and (iii) low dimensionality. J is simply proportional to the band width of energy dispersion. For that a large J means fast velocity. Also, a small J' prevents spins from scattering by the next spin network and promotes the thermal current. We arrive at the conclusion that materials in which spins are connected strongly to each other and are fluctuating are adopted for realizing thermal conductivity by spins.

Exercise

1.1 According to the kinetic theory of gases, the thermal conductivity is given by $\kappa = \lambda v c_v/3$, where λ, v, and c_v are the mean free path, the velocity, and the specific heat at constant volume. For an electron system, $c_v = (\pi^2 n k_B^2)(T/E_F)$ and the Fermi energy E_F in c_v is $m v_F^2/2$, where n and k_B are, respectively, the number density and the Boltzmann constant. Electrical conductivity is, on the other hand, expressed as $\sigma = ne^2\tau/m$ using the relaxation time τ. Derive the Wiedemann–Franz law by using the above equations.

References

1. Jahn H, Teller E (1937), Stability of polyatomic molecules in degenerate electronic states. I. Orbital degeneracy, *Proc. R. Soc. London, Series A, Math. Phys. Sci.* **161**(905), 1934–1990.

2. Zaanen J, Sawatzky GA, Allen JW (1985), Band gaps and electronic structure of transition-metal compounds, *Phys. Rev. Lett.*, **55**, 418–421.

3. Ziman JM (1964), *Principles of the Theory of Solids*, Cambridge University Press, Cambridge, UK.

4. Georges A, Kotliar G, Krauth W, Rozenberg MJ (1996), Dynamical mean-field theory of strongly correlated fermion systems and the limit of infinite dimensions, *Rev. Mod. Phys.*, **68**, 13–125.

5. Zhang XY, Rozenberg MJ, Kotliar G (1993), Mott transition in the $d = \infty$ Hubbard model at zero temperature, *Phys. Rev. Lett.*, **70**, 1666–1669.

6. Chaikin PM, Beni G (1976), Thermopower in the correlated hopping regime, *Phys. Rev. B*, **50**, 647–651.

7. Heikes RR, Ure RW, Jr. (1961), *Thermoelectricity: Science and Engineering*, Interscience, New York, London.

8. Koshibae W, Tsutsui K, Maekawa S (2000), Thermopower in cobalt oxides, *Phys. Rev. B*, **62**, 6869–6872.

9. Koshibae W, Maekawa S (2001), Effects of spin and orbital degeneracy on the thermopower of strongly correlated systems, *Phys. Rev. Lett.*, **87**, 236603-1–236603-4.

10. Wiedemann GH, Franz R (1853), Ueber die warme-leitungsfahigkeit der metalle, *Ann. Phys. Chem.*, **89**, 457–531.

11. Drude P (1900), On the electron theory of metals, *Ann. Phys.*, **1**, 566–613.

12. De Gennes PG (1958), Inelastic magnetic scattering of neutrons at high temperatures, *J. Phys. Chem. Solids*, **4**, 223–226.

13. Huber DL, Semura JS (1969), Spin and energy transport in anisotropic magnetic chains with $S = 1/2$, *Phys. Rev.*, **182**, 602–603.

14. Kudo K, Ishikawa S, Noji T, Adachi T, Koike Y, Maki K, Tsuji S, Kumagai K (1999), Spin gap and hole paring of $Sr_{14-x}A_xCu_{24}O_{41}$ (A = Ca and La) single crystal studied by electrical resistivity and thermal conductivity, *J. Low Temp. Phys.*, **117**, 1689–1693.

15. Sologubenko AV, Felder E, Gianno K, Ott HR, Vietkine A, Revcolevschi A (2000), Thermal conductivity and specific heat of the linear chain cuprate Sr_2CuO_3: evidence for thermal transport via spinons, *Phys. Rev. B*, **62**, R6108–R6111.

Chapter 2

Spin-State Crossover

2.1 Prologue

Cobalt oxides have attracted considerable attention because of their peculiar and rich physical/chemical properties such as a spin-state crossover, huge thermoelectric power, high energy density as Li ion batteries, and superconductivity, partly resulting from the simultaneous existence of strong electron–electron interaction within the Co ion and a competing hopping energy between Co $3d$ and O $2p$ states, and a small difference in the total energies for spin states. Here, the concept of the valence and spin state of Co ions plays a key role and is introduced in Section 2.2. Their rich variety is generally believed to be associated with the internal degree of freedom of the electrons. In Section 2.3, I would like to focus attention on the spin-state crossover, in which the way of occupying orbitals for the spins is changed, whereas the number of electrons remains constant. Furthermore, typical spin-state crossover systems such as $RCoO_3$ (R: the rare-earth element), $[Ca_2CoO_3][CoO_2]_{1.62}$, and $Ba_2Co_9O_{14}$ are also presented. The spin-state crossover could lead to a change in the density of states near the Fermi level, resulting in unconventional transport properties. The metal–insulator transition serves as an example, which is shown

Functional Cobalt Oxides: Fundamentals, Properties, and Applications
Tsuyoshi Takami
Copyright © 2014 Pan Stanford Publishing Pte. Ltd.
ISBN 978-981-4463-32-4 (Hardcover), 978-981-4463-33-1 (eBook)
www.panstanford.com

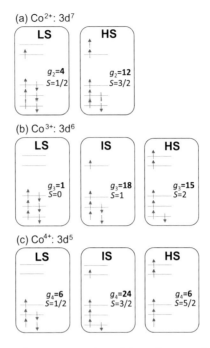

Figure 2.1 Schematic representation of the valence and spin state for (a) Co^{2+}, (b) Co^{3+}, and (c) Co^{4+}. The lower three lines and upper two lines indicate the energy level of degenerate t_{2g} and e_g orbitals, respectively. The arrow represents a spin. The degeneracy of each state (g_3 and g_4) is denoted by bold font. LS, IS, and HS represent the low-spin, intermediate-spin, and high-spin states, respectively.

in Section 2.4. Here is another example in Section 2.5: anomaly in thermal conductivity due to the lattice contribution.

2.2 Valence and Spin State of Co Ions

In general, the plural valences of +2, +3, and +4 are actually realized for Co ions; under these valence states, three spin configurations are possible, that is, low-spin (LS), intermediate-spin (IS), and high-spin (HS) states except Co^{2+} with the LS or HS state. Co $3d$ orbitals originally fivefold degenerate split into threefold t_{2g} orbitals and twofold e_g orbitals in the ideal CoO_6 octahedron. For

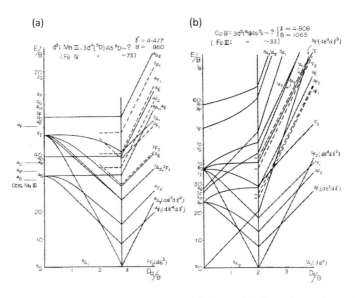

Figure 2.2 Sugano–Tanabe diagram for (a) d^5 and (b) d^6 configurations [1].

instance, if Co ion is trivalent, six electrons are placed in these orbitals. The way of spin configuration decides the spin state: $t_{2g}{}^6 e_g{}^0$, $t_{2g}{}^5 e_g{}^1$, and $t_{2g}{}^4 e_g{}^2$, respectively, correspond to the LS, IS, and HS states of Co^{3+} (see Fig. 2.1b). The spin configuration, which makes the spin angular momentum maximum, leads to the HS state ($S = 2$) according to Hund's rule. When the crystal field splitting is larger than the Hund coupling, all the spins occupy lower t_{2g} orbitals, namely, the LS state ($S = 0$). Another spin configuration is the IS state ($S = 1$).

Figure 2.2 shows a famous Sugano–Tanabe diagram calculated by using Racah's and the crystal field parameters [1]. The Coulomb integral and exchange integral are described by three Racah's parameters A–C. Among them, one gives all the electronic configurations a constant energy, while the rest two have a proportional relationship. Thus, the relative energy level can be expressed by the crystal field parameter (D_q) and one Racah's parameter, say, B. For the d^5 and d^6 cases, the HS state ($^6A_1$, $^5F_2$) is stable for small D_q/B, whereas the LS state ($^2F_2$, $^1A_1$) is stable for larger D_q/B. The IS state is observable only for this case against d^7 configuration (Co^{2+}); they

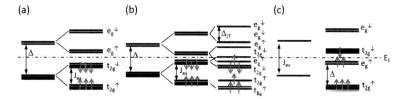

Figure 2.3 Schematic diagram of the trivalent Co ions with (a) the LS, (b) IS, and (c) HS states. Δ, J_{ex}, and E_F represent the crystal field energy, the intratomic exchange energy, and the Fermi energy, respectively.

correspond to $^4T_1$ and $^3T_1$, respectively. However, this state never becomes a ground state in Fig. 2.2. The existence of IS Co^{3+} or IS Co^{4+} implies the wave function of $3d$ electron cannot be approximated by atomic $3d$ orbitals. Although this state is normally energetically unstable, lowering the symmetry of the crystal field makes this state stable by the Jahn–Teller effect [2]. The degeneracy for each spin state is shown as g by the bold font in Fig. 2.1 (see Exercise 2.1). The point that these states are energetically close due to a competition between Hund's rule coupling and the crystal field splitting deserves explicit emphasis.

To make clear the difference between these spin states, the schematic diagram is displayed in Fig. 2.3. Both t_{2g} and e_g orbitals split into bonding and antibonding orbitals. When the crystal field energy Δ is larger than the intratomic exchange energy J_{ex}, all the spins are delivered into t_{2g} orbitals (see Fig. 2.3a). The lattice distortion might cause further splitting of orbitals. In Fig. 2.4, the relationship between the wave function and the normalized lattice constant is described. Two orbitals are found to degenerate for $c/a = 1$, but they split for $c/a = \alpha$. If one electron is delivered, it occupies the orbital allocated by φ_y, apparently resulting in a decrease in the total energy. In Fig. 2.3b, both t_{2g} and e_g orbitals split by the same effect, that is, the Jahn–Teller effect by Δ_{JT}. In contrast to the LS state, there is a spin seated in the e_g orbital, and eventually the IS state is realized. For $\Delta < J_{ex}$, on the other hand, spins are distributed from the bottom level $t_{2g}^\uparrow$ to the $t_{2g}^\downarrow$ level obeying Hund's rule, as designated in Fig. 2.3c.

Hubbard proposed a model in which both the extended electronic states based on band theory and the localized states

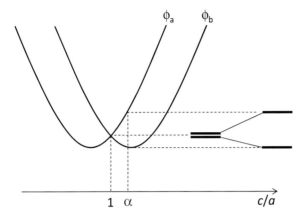

Figure 2.4 Relationship between the wave function and the normalized lattice constant, c/a.

dominated by the on-site Coulomb energy are treated [19]. The Hubbard Hamiltonian is expressed as

$$\mathcal{H} = -t \sum_{ij} (c_{i\uparrow}^+ c_{j\uparrow} + c_{j\downarrow}^+ c_{j\uparrow}) + U \sum_i n_{i\uparrow} n_{i\downarrow}, \qquad (2.1)$$

where $c_{i\uparrow}^+$ ($c_{i\uparrow}$) creates (annihilates) a spin-up electron at site i and $n_{i\uparrow}$ ($n_{i\downarrow}$) denotes the number of spin-up (down) electrons at site i ($n_{i\uparrow}$, $n_{i\downarrow}$ = 1 or 0). For $t = 0$, $\mathcal{H}$ means the isolated atomic assembly because of no interatomic transfer. When the $3d$ orbital locates below the Fermi level, the situation where spin-up and spin-down electrons occupy at site i is energetically stable for $U = 0$, whereas spin-up or spin-down electrons distribution is realized for $U > 0$. Reflection on some of these will make clear that these configurations correspond to the LS and HS states, respectively.

2.3 Spin-State Crossover

Changing the magnitude of spins of metallic atoms in materials can be regarded as a minimum magnetic switch. This phenomenon is called the spin-state transition or the spin-state crossover, in which the spin configuration responds to various external perturbations such as temperature, magnetic field, pressure, and light. This has

been one of the most basic physical properties, with thousands of examples now recognized since the discovery on tris-dithiocarbabates
of Fe^{3+} in the early 1930s. The topic that one is familiar with is
the transmission of oxygen, which becomes easy by the spin-state
transition of Fe ions in hemoglobin. It has been also pointed out
that the spin-state transition of Fe ion inside of the mantle will
attributable to an abnormal velocity of earthquake waves. Co far as
the Co ion is concerned, $[Co(PdAdH)_2]I_2$ is the first example [4, 5].

LaCoO$_3$ has been a well-known compound since the 1950s, and is
a case in point. As a famous crystal structure among transition metal
oxides, the perovskite structure, possessing the chemical formula of
AMO_3 (A: the rare-earth element and/or the alkaline-earth metal,
M: the transition metal) crosses one's mind. AMO_3 corresponds
to $n = \infty$ in the Ruddlesden–Popper family $A_{n+1}M_nO_{3n+1}$. LaCoO$_3$
also has this type of structure with $A =$ La and $M =$ Co. The
discovery of ferroelectricity in BaTiO$_3$ made the research on the
perovskite structure start [6]. Figure 2.5a shows the most basic
crystal structure called simple perovskite. An M atom is surrounded
by an octahedron made of six oxygen atoms. The edge-sharing MO_6
octahedra spread over three dimensionally.

Particular attention has been given to the spin-state transition
with increasing temperature from the nonmagnetic LS state to
higher spin states (IS and/or HS states) of the trivalent cobalt ion
Co^{3+},[7] which manifests itself in anomalies in the temperature
dependences of electrical resistivity, magnetic susceptibility, nuclear
magnetic resonance (NMR) signal, etc., at around 100 K. Central
to this issue is the problem of a precise spin state after the spin-
state crossover; the thermally driven spin-state crossover can be
confirmed from Fig. 2.6 [8]. The finite χ below 30 K despite an
LS Co^{3+} reflects the retention of tiny magnetic Co ions, but this
is irrelevant to the main subject. $\chi(\chi^{-1})$ increases (decreases)
gradually with temperature in the range of 30–100 K due to a
spin-state crossover. Above 100 K, $\chi(\chi^{-1})$ decreases (increases)
by thermal fluctuations, obeying the Curie–Weiss law. It has still
not established whether this transition is from an LS to an HS
state, to an IS state, or to their mixed state. Around 500 K where
the slope of $\chi(T)$ curve changed as well, the metal–insulator
transition takes place (see next section). Maris et al. gave evidence

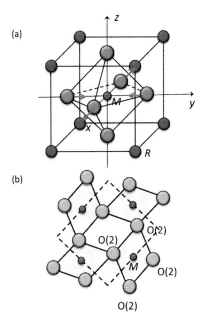

Figure 2.5 Simple perovskite structure, in which an MO_6 octahedron is formed. In (a), the arrows show the change of position due to the Jahn–Teller distortion observed for LaCoO$_3$. Its *c*-plane view is displayed in (b) as an unit cell.

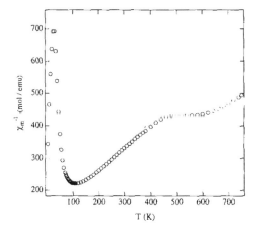

Figure 2.6 Reciprocal magnetic susceptibility vs. temperature for LaCoO$_3$ [8].

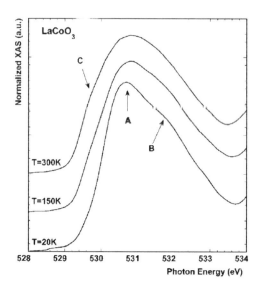

Figure 2.7 Temperature dependences of the OK-edge prepeak measured at 20 K, 150 K, and 300 K for $LaCoO_3$ [10].

of thermally excited IS Co^{3+} due to its partially filled e_g orbitals by skillfully performing the precise crystal structure analysis focusing the JahnTeller active IS Co^{3+} [9]. The Jahn–Teller distortion actually showed a steep change around the spin-state transition temperature (see Fig. 2.5b). The spin-state crossover is expected to influence the actual electronic structure. Figure 2.7 shows the OK-edge prepeak at temperatures before and after the crossover. At 20 K, double peak structures A and B are observed, and then a peak at C newly appears, as well as the broadness above the transition temperature, indicative of lattice expansion and increased lattice vibration connected to the transition [10]. From the theoretical side, the electronic structure was calculated in one-electron band structure calculations based on the local density approximation and a Coulomb correlation, so the total energies for the LS and IS states compete and reserve around 100 K (see Fig. 2.8) [11]. Please recall here the relationship between energy and temperature, for example, 0.025 eV $\approx$ 300 K.

On the other hand, the substitution of La by other rare-earth ions with smaller radius has been reported to stabilize the LS state to higher temperatures because chemical pressure increases the

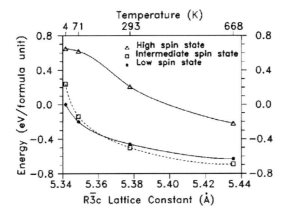

Figure 2.8 Total energies of LS, IS, and HS states of LaCoO$_3$. The upper axis shows temperature [11].

crystal field-splitting energy. At what temperature the trivalent Co ion undergoes a spin-state transition leaves room for a variety of interpretations because its temperature shows a measurement tool dependence besides an influence from the magnetic moment of R^{3+} itself, but at least it is likely that it occurs at a temperature that is higher than that of LaCoO$_3$. Figure 2.9 shows the temperature dependence of magnetic susceptibility for RCoO$_3$ (R = La, Pr, and Nd) [12]. From the main panel, the transition temperatures are assigned to be 200 K and 300 K for R = Pr and Nd, respectively. The increase and decrease in the spin-state temperature by substituting the rare-earth element with a smaller ionic radius and the alkaline earth metal, respectively, agree with the behavior of Jahn–Teller distortion [14]. In contrast to these evolutions, the alkaline-earth metal substitution unstabilizes the LS state; that is, lowers the spin-state transition temperature. Apart from chemical pressure, physical pressure makes the energy gap between the LS ground state and higher spin states increase remarkably [13].

The CoO$_6$ octahedra in LaCoO$_3$ arrays three-dimensionally by corner sharing, while edge sharing and face sharing can bring about 2D and 1D arrangements, respectively. The 2D [Ca$_2$CoO$_3$][CoO$_2$]$_{1.62}$ exhibits a spin-state transition around 420 K [15]. The crystal structure of [Ca$_2$CoO$_3$][CoO$_2$]$_{1.62}$ consists of alternating layers of the triple rock-salt-type Ca$_2$CoO$_3$ subsystem and the single

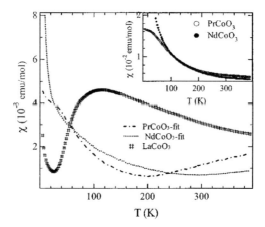

Figure 2.9 Temperature variation of magnetic susceptibility for $RCoO_3$ (R = La, Pr, and Nd) after the subtraction of the contributions from magnetic Pr^{3+} and Nd^{3+}, while the inset shows the raw data [12].

CdI_2-type CoO_2 subsystem stacked along the c axis. The corresponding $\chi(T)$ curve shows an abrupt change around 420 K as well as an anomaly on the resistivity curve, above which the effective magnetic moment is estimated to be 2.83 μ_B. The magnetic transition could then be viewed as a spin-state transition from the LS to the IS state for Co^{4+} and Co^{3+}. Associated with conductivity and magnetic properties, the Co ions in the CoO_2 planes are most likely to change their spin state. μSR measurements have a role to play in understanding the spin-state transition: both shoulders in the muon spin relaxation rate and the shift of the muon precession frequency are attributable to the spin-state transition without the magnetic order, as is evidenced by the temperature-independent asymmetry [16].

$Ba_2Co_9O_{14}$ including a 1D structure undergoes the transition around 570 K [17]. The rhombohedral $R\bar{3}m$ structure of $Ba_2Co_9O_{14}$ is shown in Fig. 2.10; five crystallographically independent Co sites (Co1–Co5) exist, and they are classified into two types of coordination polyhedra, octahedrons and tetrahedrons [18]. The $Co2O_6$ octahedron shares its opposite triangular faces with two $Co1O_6$ octahedra, forming the Co_3O_{12} octahedral trimer along the c axis; the Co4 and Co5 octahedra share edges to form the

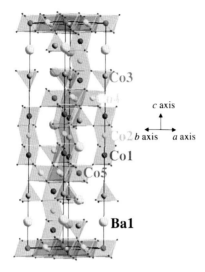

Figure 2.10 Crystal structure of $Ba_2Co_9O_{14}$.

CdI$_2$-type layer, with a 2:1 ordered arrangement within a 2D triangular lattice; and the Co3 tetrahedron shares corners with both the Co_3O_{12} trimers and the CdI$_2$ layers. All of this amounts to saying that 1D and 2D structures coexist in a material. The valence and spin-state distributions of each site were reported by Ehora et al., as listed in Table 2.1. Insulator-to-insulator phase transition at 570 K was observed, above which both the thermoelectric power and electrical resistivity decreased significantly. In addition, the $\chi^{-1}(T)$ curve deviated from the Curie–Weiss law and bent upward, which indicates a link between electrons and spins. Figure 2.11 displays the

Table 2.1 Number, valence, and spin state of Co1–Co5 in $Ba_2Co_9O_{14}$ [18]

Atom	Number	Valence	Spin state
Co1	6	+3	Low spin
Co2	3	+3	Low spin
Co3	6	+2	High spin
Co4	9	+3	Low spin
Co5	3	+2	High spin

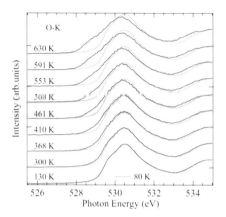

Figure 2.11 Temperature-dependent O-K XAS spectra of $Ba_2Co_9O_{14}$ together with the data taken at 80 K (green line) for comparison [17].

temperature-dependent O-K X-ray absorption spectroscopy (XAS) spectra of $Ba_2Co_9O_{14}$. The unoccupied states related to the Co^{3+} ions located at the pre-edge region below 532 eV and originated from transitions from the O $1s$ core level to the O $2p$ orbitals that are mixed into the unoccupied Co $3d$ states. For a LS Co^{3+} below 300 K with the $t_{2g}^6 e_g^0$ configuration, the lowest-energy structure in the O-K spectrum $\approx$ 530.5 eV is due to the transitions from the O1s core level into the unoccupied Co $3d$ e_g states, because the low-lying t_{2g} orbitals are fully occupied. A lower-energy spectral feature $\approx$ 528.7 eV appears and becomes pronounced above 510 K, whereas the peak at 530.5 eV loses its spectral weight, which indicates the t_{2g} states are no longer fully occupied. To isolate the site where the spin-state transition takes place, the average lengths is estimated. Figure 2.12 captures that the bond lengths of Co3–O, Co4–O, and Co5–O show almost no change or a slight decrease; and Co1–O and Co2–O in the octahedral trimer expand dramatically in the vicinity of 570 K. The temperature dependence of the bond length in the trimer in the measured temperature range is out of the possible range of the thermal expansion in a typical perovskite oxide. Thus, we are now able to see that the trivalent Co1 and Co2 ions within the octahedral trimer are responsible for the spin-state transition. Their transition temperature variation of dimensionality seems to

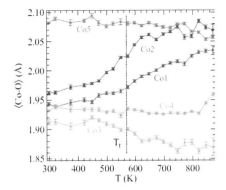

Figure 2.12 Temperature dependences of the average Co–O bond lengths on different five sites Co1–Co5 in $Ba_2Co_9O_{14}$ [18]. T_t shows the insulator-to-insulator transition temperature.

be increased with decreasing the dimensionality. Recently, the spin-state transition was discovered also for a four-ligand system, $SrFeO_2$ [19].

2.4 Metal-Insulator Transition

Wave particle duality for a strongly correlated electron system is notably effective as a phase transition between localized and itinerant phases. One could say that unconventional physical properties occur at the boundary between the insulator and the metal. In general, to make the Mott insulator metalization, whether changing the band width or the band filling by doping carriers is required. They are so-called the band width control or filling control, respectively. Regardless of the kind of insulators, the spin-state transition could cause a change in the electronic structure near the Fermi level and, in some cases, results in a metal–insulator transition (MIT). Although heretofore we discussed mainly the spin-state transition around 100 K for $LaCoO_3$, this material also shows an MIT around 500 K. With an increase of temperature, the electrical resistivity and thermoelectric power change from a semiconducting state to a fairly conducting state, in addition to a gradual change in χ around this temperature, as is standard in the spin gap that arises perhaps from the level difference between the LS and HS states of the Co^{3+} ion.

The MIT also results from the overlapping of the metallic spin cluster that emerges from the close proximity of multiple randomly doped Sr ions in $R_{1-x}Sr_xCoO_3$. For instance, when $R = $ La, the system simultaneously undergoes a percolative MIT and a ferromagnetic transition at the critical doping of $x_c = 0.17$. An ordinary MIT observed for $R_{1-x}Sr_xCoO_3$ comes from the change in the electronic structure due to the spin-state crossover, but as an interesting phenomenon, we take up $Pr_{0.5}Ca_{0.5}CoO_3$, which exhibits an MIT at 70 K ($= T_{MIT}$). At the beginning, charge ordering between LS Co^{3+} and IS Co^{4+} is formed below T_{MIT} and holes of IS Co^{4+} embedded in IS Co^{3+} produced *via* the spin-state crossover [20]. However, recently the charge immigration effect of $Pr^{3+/4+}$ on the MIT has been argued: Co ions whose average valence state is 3.5 are deoxidized by receiving electrons from Pr^{3+}; that is, $\frac{1}{2}Pr^{3+} + Co^{3.5} \rightarrow \frac{1}{2}Pr^{3+2\delta} + Co^{3.5-\delta}$. This type of MIT coming from an increase in the number of electrons for Co ions, being different from the electron configuration effect on orbitals, that is, the spin-state crossover, cannot be overemphasized [21].

Beside these MIT transitions rise from many-body effect between electrons, there exists those discussed in the range of one-electron approximation: (i) Peierls transition, (ii) Anderson transition, (iii) Bloch-Wilson transition, and (iv) percolation transition. For the Peierls transition, due to spontaneous periodic deformation, the gap appears at the Fermi level, resulting in the insulator as first reported for $CuGeO_3$ among inorganic materials [22]. The Anderson transition is attributable to the localization of electrons by irregular potentials. Mechanism of Bloch-Wilson transition is the intersection of band structure by changing pressure and/or temperature. For the percolation scenario, ratio between the metallic and insulating phases is a key to occur the metal-insulator transition.

2.5 Thermal Rectifier

Figure 2.13 shows the temperature dependence of thermal conductivity for $La_{1-x}Sr_xCoO_3$. The thermal conductivity (κ) of $LaCoO_3$

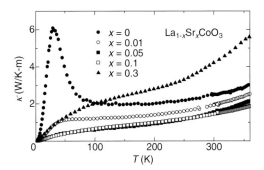

Figure 2.13 Temperature dependence of thermal conductivity for $La_{1-x}Sr_xCoO_3$.

shows a sharp peak around 20 K and drops faster than a $1/T$ law, which reflects the phonon contribution due to its insulating behavior. Thus, the spin-state crossover suppresses the phonon thermal conductivity *via* lattice distortion. With a slight doping of Sr, κ shows a decay and its value is comparable to that of the $NaCo_2O_4$ polycrystalline sample. Further introduction of carriers brings an increase in κ since the electronic thermal conductivity increases, reflecting good conductivity.

This nature can be applicable as the thermal rectifier. When the materials A and B are connected each other, material A exhibits a high κ at low temperature (T_c) and a lower κ at high temperature (T_h), and material B behaves vice versa. Namely, heat flows in a forward direction from material B at T_h and to A at T_c and hardly when T_h and T_c are switched, which is regarded as a kind of diode. Recently, an oxide thermal rectifier composed of $LaCoO_3$ (material A) and $La_{0.7}Sr_{0.3}CoO_3$ (material B) was prepared, and its phenomenon was demonstrated [23].

Exercise

2.1 Derive the total spin and orbital degeneracy of Co ions shown in Fig. 2.1.

References

1. Tanabe Y, Sugano S (1954), On the absorption spectra of complex ions II, *J. Phys. Soc. Jpn.*, **9**, 766–779.

2. Jahn H, Teller E (1937), Stability of polyatomic molecules in degenerate electronic states. I. Orbital degeneracy, *Proc. R. Soc. Lond., Series A, Math. Phys. Sci.*, **161**(905), 1934–1990.

3. Hubbard J (1963), Electron correlations in narrow energy bands, *Proc. R. Soc. Lond.*, **A276**, 238–257.

4. Cambi L, Szego L (1931), Uber die magnetische susceptibilitat der komplexen verbindungen, *Chem. Ber. Dtsch. Ges.*, **64**, 2591–2597.

5. Stoufer RC, Busch DH, Hadley WB (1961), Unusual magnetic properties of some six-coordinate cobalt(II) complexes' electronic isomers, *J. Am. Chem. Soc.*, **83**, 3732–3734.

6. Wainer E, Salomon AN (1942), *Titanium Alloy Mfg. Co. Elec. Rep.*, **8**.

7. Goodenough JB (1958), An interpretation of the magnetic properties of the perovskite-type mixed crystals $La_{1-x}Sr_xCoO_{3-\lambda}$, *J. Phys. Chem. Solids*, **6**, 287–297.

8. Señarís-Rodríguez MA, Goodenough JB (1980), $LaCoO_3$ Revisited, *J. Solid State Chem.*, **116**, 224–231.

9. Maris G, Ren Y, Volotchaev V, Zobel C, Lorenz T, Palstra TTM (2003), Evidence for orbital ordering in $LaCoO_3$, *Phys. Rev. B*, **67**, 224423-1–224423-5.

10. Toulemonde O, N'Guyen N, Studer F, Traverse A (2001), Spin state transition in $LaCoO_3$ with temperature or strontium doping as seen by XAS, *J. Solid State Chem.*, **158**, 208–217.

11. Korotin MA, Ezhov SY, Solovyev IV, Anisimov VI, Khomskii DI, Sawatzky GA (1996), Intermediate-spin state and properties of $LaCoO_3$, *Phys. Rev. B*, **54**, 5309–5316.

12. Yan JQ, Zhou JS, Goodenough JB (2004), Bond-length fluctuations and the spin-state transition in $LCoO_3$ ($L =$ La, Pr, and Nd), *Phys. Rev. B*, **69**, 134409-1–134409-6.

13. Asai K, Yokokura O, Suzuki M, Naka T, Matsumoto T, Takahashi H, Môri N, Kohn K (1997), Pressure dependence of the 100 K spin-state transition in $LaCoO_3$, *J. Phys. Soc. Jpn.*, **66**, 967–970.

14. Takami T, Zhou JS, Goodenough JB, Ikuta H (2007), Correlation between the structure and the spin state in $R_{1-x}Sr_xCoO_3$ ($R =$ La, Pr, and Nd), *Phys. Rev. B*, **76**, 144116-1–144116-7.

15. Masset AC, Michel C, Maignan A, Hervieu M, Toulemonde O, Studer F, Raveau B (2000), Misfit-layered cobaltite with an anisotropic giant magnetoresistance: $Ca_3Co_4O_9$, *Phys. Rev. B*, **62**, 166–175.

16. Sugiyama J, Brewer JH, Ansaldo EJ, Itahara H, Dohmae K, Seno Y, Xia C, Tani T (2003), Hidden magnetic transitions in the thermoelectric layered cobaltite $[Ca_2CoO_3]_{0.62}CoO_2$, *Phys. Rev. B*, **68**, 134423-1–134423-8.

17. Cheng JG, Zhou JS, Hu Z, Suchomel MR, Chin YY, Kuo CY, Lin HJ, Chen JM, Pi DW, Chen CT, Takami T, Tjeng LH, Goodenough JB (2012), Spin-state transition in $Ba_2Co_9O_{14}$, *Phys. Rev. B*, **85**, 094424-1–094424-6.

18. Ehora G, Daviero-Minaud S, Colmont M, André G, Mentré O (2007), $Ba_2Co_9O_{14}$: New inorganic building blocks with magnetic ordering through super-super exchanges only, *Chem. Mater.*, **19**, 2180–2188.

19. Kawakami T, Tsujimoto Y, Kageyama H, Chen XQ, Fu CL, Tassel C, Kitada A, Suto S, Hirama K, Sekiya Y, Makino Y, Okada T, Yagi T, Hayashi N, Yoshimura K, Nasu S, Podloucky R, Takano M (2009), Spin transition in a four-coordinate iron oxide, *Nat. Chem.*, **1**, 371–376.

20. Tsubouchi S, Kyomen T, Itoh M, Ganguly P, Oguni M, Shimojo Y, Morii Y, Ishii Y (2002), Simultaneous metal-insulator and spin-state transitions in $Pr_{0.5}Ca_{0.5}CoO_3$, *Phys. Rev. B*, **66**, 052418-1–052418-4.

21. García-Muñoz JL, Frontera C, Barón-González AJ, Valencia S, Blasco J, Feyerherm R, Dudzik E, Abrudan R, Radu F (2011), Valence transition in (Pr,Ca)CoO_3 cobaltites: charge migration at the metal-insulator transition, *Phys. Rev. B*, **84**, 045104-1–045104-6.

22. Hase M, Terasaki I, Uchinokura K, (1993), Observation of the spin-Peierls transition in linear Cu^{2+} (spin-1/2) chains in an inorganic compound $CuGeO_3$, *Phys. Rev. Lett.*, **70**, 3651–3654.

23. Kobayashi W, Teraoka Y, Terasaki I (2009), An oxide thermal rectifier, *Appl. Phys. Lett.*, **95**, 171905-1–171905-3.

Chapter 3

Li Ion Battery

3.1 Prologue

Batteries can be defined as "devices to produce directly the chemical, physical, and biological change of composed materials as the electrical energy." They fall into three kinds, that is, chemical, physical, and biological batteries, as in Fig. 3.1. A chemical battery can produce electrical energy by a chemical reaction in the cell. A physical battery can convert light and thermal energies into electrical energy without a chemical reaction. A biological battery can produce electrical energy by a biochemical change using a biocatalyst and a microorganism. Among them, we usually call chemical batteries such as a dry cell battery and a storage battery used in social life as batteries. Added to this, chemical batteries are grouped into a dischargeable primary battery, a charge/dischargeable second battery, and a fuel cell that can be a future electronic source. Cobalt oxides are actively involved in a second battery and a fuel cell, which are introduced in this chapter and Chapter 10, respectively.

Rechargeable lithium ion batteries are light and have a high capacitance and a high electromotive force compared to conventional lead-acid storage batteries and rechargeable nickel-cadmium

Functional Cobalt Oxides: Fundamentals, Properties, and Applications
Tsuyoshi Takami
Copyright © 2014 Pan Stanford Publishing Pte. Ltd.
ISBN 978-981-4463-32-4 (Hardcover), 978-981-4463-33-1 (eBook)
www.panstanford.com

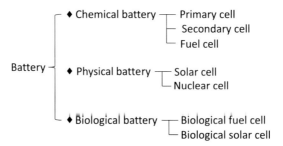

Figure 3.1 Kinds of batteries.

batteries. Nowadays when the mobile apparatus has diversified and has had higher functionality, the role of rechargeable lithium ion batteries becomes more and more essential. Rechargeable batteries, which consist of three parts, that is, a positive electrode, an electrolyte solution, and a negative electrode, can charge or discharge electrical energy converted from chemical energy by an oxidation-reduction reaction at both electrodes. For rechargeable lithium ion batteries, Li that is used as a carrier of electrical energy involves conduction in the route from a positive electrode to a negative electrode via an electrolyte solution. $LiCoO_2$ has still been the mainstream as a positive electrode since its discovery in 1980 [1]. The history of batteries is reviewed in chronological order in Section 3.2, and then details of $LiCoO_2$ are specified in Section 3.3.

3.2 History of Batteries

In 1700s, L. Galvani demonstrated what we now understand to be the electrical basis of nerve impulses, which yielded the cornerstone of research by later inventors to create batteries. In early 1800s, investigation of the batteries—the voltaic pile, Daniell cell, Poggendorff cell, and Grove cell—was actively performed. The voltaic pile discovered by A. Volta, can produce an electrical current by alternating discs of zinc and copper with pieces of cardboard soaked in brine between the metals. The metallic conducting arc was used to carry electricity over a greater distance. Volta's voltaic pile was the first wet cell battery that produces a reliable and steady current of electricity. With all these advantages, the voltaic pile could not deliver the electrical current for a long period of time. J. F. Daniell, on the other hand, invented a cell consisting of two electrolytes, copper and zinc sulfates, which is called the Daniell cell. This popular battery, which produces approximately 1.1 V, was widely used to power objects such as telegraphs, telephones, and doorbells.

J. C. Poggendorff technically overcame the ways with separating the electrolyte and the depolarizer using a porous earthenware pot. For the Poggendorff cell, the electrolyte and the depolarizer were, respectively, dilute sulfuric and chromic acid. These acids were physically mixed together, eliminating the porous pot. The positive electrode (cathode) was two carbon plates, while the anode was a zinc plate. It has been known for many years due to its relatively high voltage of 1.9 V goes beyond the Daniell cell, greater ability to produce a consistent current, and lack of any fumes, but the relative fragility of its thin glass enclosure and the necessity of having to raise the zinc plate when the cell was not in use. The cell was simply regarded as the chromic acid cell but principally as the bichromate cell. The latter name came from the practice of producing chromic acid by adding sulfuric acid to potassium bichromate, even though the cell itself contained no bichromate. The Grove cell, explored by W. Robert Grove in 1844, was composed of a zinc anode dipped in sulfuric acid and a platinum cathode dipped in nitric acid, separated by porous earthenware. The Grove cell had advantages of a high current and nearly two times larger than the voltage of the Daniell cell, with disadvantages of poisonous nitric acid during operation, a sharp drop in voltage as the charge diminished, and rare and expensive platinum. The Grove cell came to be improved by the cheaper, safer, and better-performing gravity cell in the 1860s.

In the late 1800s, rechargeable batteries and dry cells including lead-acid cells, gravity cells, carbon-zinc cells, and nickel-cadmium cells, were reported. G. Plante developed the first practical storage lead-acid battery that could be; that is, one can regard it as a secondary battery. This type of battery is primarily used in cars today. G. Leclanche patented the carbon-zinc wet cell battery called the Leclanche cell in 1866. This battery can supply a voltage of 1.4–1.6 V as well as an electrical current for a long time. The positive electrode consisted of crushed manganese dioxide with a tiny carbon mixed in, while the negative pole was a zinc rod. They are mediated by an ammonium solution, which works as the electrolyte, readily seeping through the porous cup and making contact with the cathode material.

In the 1900s, new technology and ubiquity were being further developed. To take simple examinations, nickel-iron, alkaline batteries, nickel-hydrogen and nickel metal hydride, and lithium-ion batteries were the creatures of the day. T. A. Edison invented the alkaline storage battery, whose cell had iron as the anode material and nickelic oxide as the cathode material. G. Pearson, C. Fuller, and D. Chapin performed research on the first solar battery in 1954, which converts the sun's natural energy into electricity. They created an array of several strips of silicon in the size of a razorblade. They placed them in sunlight and succeeded in turning the captured free electrons into an electrical current. The first public service trial of the Bell solar battery began with a telephone carrier system on October 4, 1955. Three important developments were marked in the 1980s. In 1980, J. B. Goodenough disclosed the $LiCoO_2$ cathode and R. Yazami discovered the graphite anode. These findings led a research team managed by A. Yoshino of Asahi Chemical, Japan, to build the first lithium ion battery prototype that was a rechargeable and more stable version of the lithium battery in 1985.

3.3 $LiCoO_2$ Batteries

Lithium cobalt oxide formulated as $LiCoO_2$ is a chemical compound commonly used as a positive electrode in lithium ion second batteries. The crystal structure of $LiCoO_2$ was originally reported by Johnston et al. [4]. As shown in Fig. 3.2, the two-dimensional CoO_2 layers consisting of edge-sharing CoO_6 octahedra are formed in the ab plane, between which Li is sandwiched with full occupation. $LiCoO_2$ with an α-$NaFeO_2$-type structure is clarified to be divided into two categories in the structural sense, that is to say, the high-temperature (HT) phase and the low-temperature (LT) phase, depending on the preparation temperature [2, 3]. The HT phase prepared at 1073 K has a layered structure with the lattice

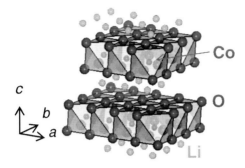

Figure 3.2 Crystal structure of LiCoO$_2$.

constants of $a = 2.817$ Å and $c = 14.058$ Å and acts as the positive electrodes. On the other hand, the electrochemical properties of the LT phase are relatively poor: unstable charge/discharge and bad cycle characteristics are notable examples.

Open-circuit voltages of 4–5 V for Li$_x$CoO$_2$/Li cells are consistent with the oxidizing power of the Co^{4+}/Co^{3+} couple. Preliminary voltage–composition curves show low overvoltages and good reversibility for current densities up to 4 mAcm^{-2} over a wide range of x, indicative of being promising for practical application. To such chemical interest, we may add the comment that the ionic conductivity and various magnetic properties for Li$_x$CoO$_2$ also drive us into further research. For stoichiometric LiCoO$_2$, Co^{3+} ions are in a low spin state with $S = 0$ at ambient temperature, resulting in a nonmagnetic band insulator. The Li deficiency means hole doping in the triple degenerate t_{2g} orbitals with decreasing x in Li$_x$CoO$_2$. To put it another way, the population of Co^{4+} ions is increased. With Li removal for $x = 0.55$–0.7, metallic behavior is observed via an insulator–metal transition around $x = 0.95$ [5]. Moreover, the thermoelectric power at room temperature for Li$_x$CoO$_2$ decreases systematically from 190 µV/K for $x = 1$ to 40 µV/K for $x = 0.6$ [5]. These behaviors have justice in the view where carriers are actually introduced as holes by Li removal.

Figure 3.3 shows the crystal structure of graphene [6]. C atoms ordinary form a hexagonal unit with the C–C bond length of 1.42 Å, which is stacked along the b axis in the interval of 3.35 Å. It is easy to detach due to weak interlayer bonding. As an allotrope, diamond,

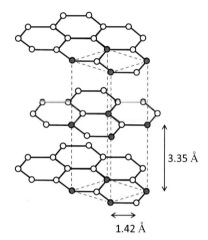

Figure 3.3 Crystal structure of graphene.

fullerene, and carbon nanotubes are well known. Quite recently, graphite treated with pure water, as Scheike et al. have reported, shows room temperature superconductivity [7].

Li ion second batteries have widely spread over the world as the electrical source of the portable phone and notebook computer since the 1990s because of their high energy density, light weight, and long cycle life. For this battery, $LiCoO_2$ has been used as the positive electrode and carbon as the negative electrode, together

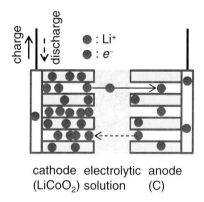

cathode   electrolytic   anode
($LiCoO_2$)   solution   (C)

Figure 3.4 Schematic illustration when the Li ion second battery operates.

with an organic electrolytic solution (see Fig. 3.4). Although there were several research in which the Li metal removal was done at the anode, the problem related to irreversibility and reactivity was still left behind due to the extremely high chemical activity of Li. The chemical reactions occur at both electrodes and are expressed as

$$LiCoO_2 \rightleftarrows Li_{1-x}CoO_2 + xLi^+ + xe^- \qquad (3.1)$$

and

$$6C + xLi^+ + xe^- \rightleftarrows Li_xC_6. \qquad (3.2)$$

Hence, the reaction of

$$LiCoO_2 + 6C \rightleftarrows Li_{1-x}CoO_2 + Li_xC_6. \qquad (3.3)$$

undergoes as a battery.

We can understand by Fig. 3.4 how the Li ion second battery operates. For the case of charge, Li^+ in Li_xCoO_2 of the positive electrode moves to interlayers of graphite through the electrolytic solution that mediates the exchange of Li ions between the anode and the cathode. Since Li removal makes the valence of Co ions increase, the transition metal ions for oxidation and reduction are required as composed elements. Li^+ moves to the opposite direction on discharging. The main characteristics of a Li ion second battery are summarized below:

- High operating voltage (4–5 V)
- High energy density (200 Wh/kg)
- Large capacitance (2.4 Ah)
- Excellent cycle characteristics
- No memory effect

Furthermore, Table 3.1 helps to take a look at the materials used in a Li ion second battery.

Table 3.1   Materials used in Li ion second batteries

Parts	Materials
Cathode	LiMO_2 (M = Co, Ni, Mn), LiMn$_2$O$_4$, Li$_x$TiS$_2$, Li$_x$V$_2$O$_5$, V$_2$MoO$_8$, MoS$_2$, LiFePO$_4$
Anode	Li, Li-Al alloy, Li-Pb alloy, SnO$_2$, Li$_{4/3}$Ti$_{5/3}$O$_4$, Li$_7$MnN, C
Electrolytic solution	LiClO$_4$, LiPF$_6$

Table 3.2 Effective nuclear charge derived from Slater's law [8]

Orbital	Sc	Ti	V	Cr	Mn	Fe	Co	N	Cu
	$(4s)^2(3d)$	$(4s)^2(3d)^2$	$(4s)^2(3d)^3$	$4s(3d)^5$	$(4s)^2(3d)^5$	$(4s)^2(3d)^6$	$(4s)^2(3d)^7$	$(4s)^2(3d)^8$	$4s(3d)^{10}$
$3s$	3.0	3.2	3.3	3.0	3.6	3.8	3.9	4.1	3.7

The storage energy of a battery is the product of the voltage (E) and the quantity of electricity (C):

$$\text{volume energy density} = EC/V \qquad (3.4)$$

and

$$\text{mass energy density} = EC/m, \qquad (3.5)$$

using the volume V and the mass m. The volume energy density, mass energy density, E, and C are often expressed in practical units of [W·hour/cm^3], [W·hour/kg], [V], and [A·hour], respectively. The theoretical capacity that is the maximum quantity of electricity is determined by the composition ratio of Li and Co. Li$^+$ is assigned to CoO$_2$ in the LiCoO$_2$ cathode and to C$_6$ in the anode. As for the second factor, provided LiM_xCo$_{1-x}$O$_2$ containing M^{3+}, only xLi$^+$ ($x < 1$) can be exchanged even though a full occupation of Li$^+$. The voltage is, on the other hand, determined by the difference between the Fermi energies (E_F) of cathode and anode materials. One should relatively lower E_F of the cathode setting E_F of the anode 0 as a dissolution/deposition voltage. The effective nuclear charge of the transition metal itself is shown in Table 3.2, which is derived on the basis of Slater's law [8]. One may say that an increase in the effective nuclear charge with the atomic number means the same effect on the binding energy as its depth. Compared to the same atom with

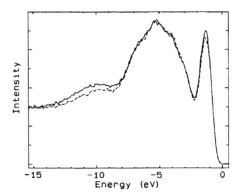

Figure 3.5 Electronic structure obtained from resonant photoemission spectroscopy for LiCoO$_2$ setting the photon energy 63 eV (solid curve) and 62 eV (dashed curve) [9].

different ionized states, a larger V can be deduced in order of Co^{4+}, Co^{3+}, and Co^{2+}. The electronic structure of $LiCoO_2$ is displayed in Fig. 3.5 [9], which tells us that the peak position is estimated to be ≈ 1.5 eV (≈ 1.5 V). To be concluded, the most important factor to control the voltage is the selection of the transition metal and its valence state.

Exercises

3.1 Check the effective nuclear charge of Co to be 3.9 using Slater's law.

3.2 Calculate the theoretical mass capacity of $LiCoO_2$ when 1 mol Li^+ can be inserted/disconnected. Then compare it with that of $LiMn_2O_4$ under the same condition.

References

1. Mizushima K, Jones PC, Wiseman PJ, Goodenough JB (1980), Li_xCoO_2 ($0 < x < 1$): a new cathode material for batteries of high energy density, *Mater. Res. Bull.*, **15**, 783–789.

2. Levasseur S, Menetrier M, Suard E, Delmas C (2000), Evidence for structural defects in non-stoichiometric HT-$LiCoO_2$: electrochemical, electronic properties and ^7Li NMR studies, *Solid State Ionics*, **128**, 11–24.

3. Gummow RJ, Liles DC, Thackeray MM, David WIF (1993), A reinvestigation of the structures of lithium-cobalt-oxides with neutron-diffraction data, *Mater. Res. Bull.*, **28**, 1177–1184.

4. Johnston WD, Heikes RR, Sestrich D (1958), The preparation, crystallography, and magnetic properties of the $Li_xCo_{(1-x)}O$ system, *J. Phys. Chem. Solids*, **7**, 1–13.

5. Motohashi T, Ono T, Sugimoto Y, Masubuchi Y, Kikkawa S, Kanno R, Karppinen M, Yamauchi H (2009), Electronic phase diagram of the layered cobalt oxide system Li_xCoO_2 ($0.0 \leqslant x \leqslant 1.0$), *Phys. Rev. B*, **80**, 165114-1–165114-9.

6. Novoselov KS, Geim AK, Morozov SV, Jiang D, Katsnelson MI, Grigorieva IV, Dubonos SV, Firsov AA (2005), Two-dimensional gas of massless Dirac fermions in graphene, *Nature*, **438**, 197–200.

7. Scheike T, Böhmann W, Esquinazi P, Barzola-Quiquia J, Ballestar A, Setzer A, (2012), Can doping graphite trigger room temperature superconductivity? Evidence for granular high-temperature supercon-ductivity in water-treated graphite powder, *Adv. Mater.*, **24**, 5826–5831.

8. Slater JC (1930), Atomic shielding constants, *Phys. Rev.*, **36**, 57–64.

9. van Elp J, Wieland JL, Eskes H, Kuiper P, Sawatzky GA, de Groot FMF, Turner TS (1991), Electronic structure of CoO, Li-doped CoO, and $LiCoO_2$, *Phys. Rev. B*, **44**, 6090–6103.

Chapter 4

Huge Thermoelectric Power

4.1 Prologue

The contribution of thermoelectric devices to energy and environ-
mental problems is expected to play a key role in the future. As
one can see from the name, thermoelectric devices can be used
for electricity generation directly from a heat source that has not
been effectively used. In this process, a physical phenomenon called
the Seebeck effect is applied. The discoverer T. J. Seebeck found
magnetic force by applying a temperature difference between both
ends of Bi being closed by a Cu wire and named it thermomagnetism.
Three years later, this phenomenon became to be recognized as
thermoelectricity. The key technology for material science has
significantly developed since 1980. In the present era, as one
might say, experience and instinct are coming to be no longer
a way to rely on. In such a stream of time, the discovery of
high-T_c cuprate superconductors gives an opportunity to go over
thermoelectric materials, too, especially for oxide materials. A great
part of this chapter is devoted to the trend of Co oxide thermoelectric
materials.

Functional Cobalt Oxides: Fundamentals, Properties, and Applications
Tsuyoshi Takami
Copyright © 2014 Pan Stanford Publishing Pte. Ltd.
ISBN 978-981-4463-32-4 (Hardcover), 978-981-4463-33-1 (eBook)
www.panstanford.com

4.2 Thermoelectric Materials

The maximum thermoelectric efficiency is yielded by the following equation:

$$\eta = \frac{\Delta T}{T_h} \frac{M - 1}{M + (T_c/T_h)}, \tag{4.1}$$

where $\Delta T/T_h$ and ΔT ($= T_h - T_c$) are the Carnot efficiency and the temperature difference between high temperature and low temperature, respectively. Furthermore, $M = (1 + ZT)^{1/2}$, which includes the dimensionless figure of merit ZT, defined as follows:

$$ZT = S^2 T/\rho \kappa, \tag{4.2}$$

where S, ρ, κ, and T are the thermoelectric power, electrical resistivity, thermal conductivity, and absolute temperature, respectively. The material that simultaneously shows a large thermoelectric power, low electrical resistivity, and low thermal conductivity is suitable for thermoelectric materials. When the thermoelectric material is linked to the battery as a simple image, this condition means that both a low internal resistance and a large voltage are required for application. A threshold for application is $ZT \approx 1$, which roughly corresponds to the thermoelectric efficiency of 10%, provided $T_c/T_h = 0.5$. Such materials are, however, rarely found because the three parameters are a function of carrier concentration in conventional interpretation so they cannot be controlled independently.

Let's formulate the situation for intuitive understanding. The free-electron model gives the relationship between the energy E and the wave number k as

$$E = \frac{\hbar^2}{2m}(k_x^2 + k_y^2 + k_z^2) = \frac{\hbar^2 k^2}{2m}. \tag{4.3}$$

The carrier concentration n is expressed as

$$n = 2 \int \frac{d^3 k}{(2\pi)^3} f(E) = \frac{1}{\pi^2} \left(\frac{2mk_B T}{\hbar^2} \right)^{3/2} \int_0^\infty \sqrt{x} \frac{dx}{e^{x-\beta\mu} + 1}, \tag{4.4}$$

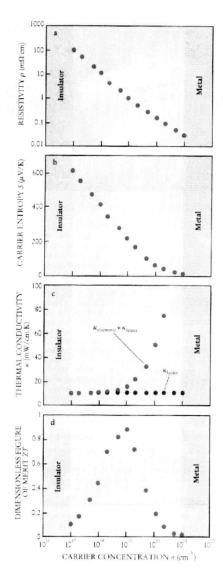

Figure 4.1 Carrier concentration dependences of (a) electrical resistivity, (b) thermoelectric power, (c) thermal conductivity, and (d) dimensionless figure of merit [2].

where $x = E/k_B T$, $\beta = 1/k_B T$, and μ is the chemical potential. For $E_F \gg k_B T$, this equation is converted to

$$n = \frac{1}{\pi^2} \left(\frac{2mk_B T}{\hbar^2} \right)^{3/2} \int_1^\infty \sqrt{x} e^{\beta \mu - x} dx = \xi e^{\beta \mu}, \qquad (4.5)$$

where ξ is the coefficient of the exponential term. From Eq. 1.59,

$$S = \frac{k_B}{e} (-\beta \mu + \delta). \qquad (4.6)$$

Thus, $\beta \mu$ is transformed to be a logarithmic form from Eq. 4.5, resulting in

$$S = -\frac{k_B}{e} \left(\log \frac{n}{\xi} - \delta \right). \qquad (4.7)$$

The Drude theory, on the other hand, suggests

$$\rho = \frac{m}{ne^2 \tau}, \qquad (4.8)$$

where m, e, and τ are the mass, the elementary charge, and the relaxation time, respectively. Furthermore, according to Eq. 1.83,

$$\kappa_e = 2.45 \times 10^{-8} \sigma T \propto n. \qquad (4.9)$$

By this means, S and ρ decrease monotonously with increasing n, while κ increases (see Fig. 4.1). For example, one tries to enhance S whereupon ρ is also increased. Then a reduction in ρ leads to an increase in κ in turn.

Figure 4.2 shows the time dependence of ZT for representative thermoelectric materials. The object of investigation changed from simple metals and half metals to intermetallic compounds around the 1950s. The Bi_2Te_3 compound was an excellent thermoelectric material for a while, and its performance reaches $ZT \approx 0.85$ at room temperature [1]. Since 1950, a material whose ZT exceeds 1 has not been found, with the exception of TAGS. In the 1990s, the materials possessing $ZT \geq 1$ were explored; all of them are degenerate semiconductors exhibiting high mobility. The guideline to designing good thermoelectric materials is simple, that is, optimization of the carrier concentration to be 10^{19} cm^{-3}, where ZT theoretically takes a maximum value when three factors are properly controlled (see Fig. 4.1d) [2]. As opposite to these alloy compounds, in 1997, a huge thermoelectric power was discovered in $NaCo_2O_4$ ($x = 0.5$

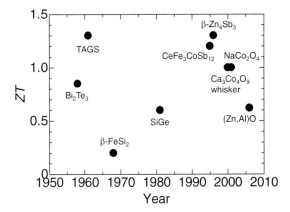

Figure 4.2 Time progress of the dimensionless figure of merit of representative thermoelectric materials. TAGS is the material composed of $AgSbTe_2$ and GeTe with a compositional ratio of 1:1.

in Na_xCoO_2) despite a low electrical resistivity comparable to the normal state of superconductors [4]. Its carrier concentration is 2 orders of magnitude higher than the optimal carrier concentration, but its ZT reaches almost 1 [5]. This breakthrough upturned the above long-held idea, and its thermoelectric properties and physics behind them are presented in the next section and Section 4.5, respectively.

4.3 Na_xCoO_2

$NaCo_2O_4$ was first synthesized by V. Jansen and R. Hopper, and its crystal structure was determined [6]. This structure is known as A_xBO_2 (0.5 < x < 1); The A site includes the alkaline metals such as Li, Na, K, Rb, and Cs, while the B site includes the $3d$ transition metals such as Mn, Fe, Co, and Ni. For the moment let us look closely at the crystal structure. In Fig. 4.3, the crystal structure of $NaCo_2O_4$ is illustrated. $NaCo_2O_4$ possesses structural components, CoO_2 and Na layers, which are stacked alternatively along the c axis. The CoO_2 layer consists of a 2D triangular lattice of edge-sharing CoO_6 octahedra in the ab plane. In the Na layer, on the other hand, Na cation is deficient by 50%, as can be checked easily from the composition ratio of $Na_{0.5}CoO_2$.

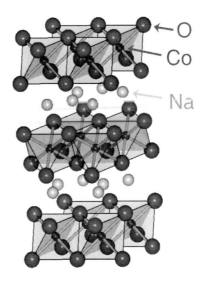

Figure 4.3   Crystal structure of $NaCo_2O_4$.

There is a previous report in which the metallic conductivity of a polycrystalline sample of $Na_{0.5}CoO_2$ remains down to 12 K [3], but Fig. 4.4 shows the temperature dependences of thermoelectric power and electrical resistivity for the $NaCo_2O_4$ single crystal [4]. The resistivity is 200 $\mu\Omega$cm at room temperature in spite of a quite low mobility besides that of conventional thermoelectric materials. A low mobility of 10 cm^2/Vs indirectly implies strong electron–phonon and/or electron–electron interactions. On the other hand, the thermoelectric power is 100 μV/K around room temperature, which is approximately a tenfold magnitude larger compared to simple metals. The thermal conductivity (κ) is also not large because the lattice contribution to κ is small due to the vacancy of Na between the CoO_2 layers, for example, 19 W/Km at 300 K, and decreases upon heating [5]. These behaviors of three parameters fulfill the condition as thermoelectric materials. The work performed by Terasaki et al. opened up a new era for the research of oxide compounds. The value of S itself, 100 μV/K, is not quite large since a much larger S is indeed achieved for the system with a low carrier concentration, such as semiconductors and insulators (see Fig. 4.1b), but a distinguishing characteristic

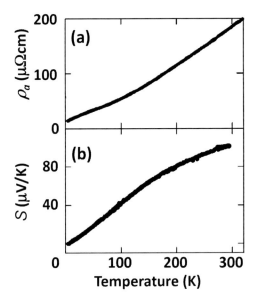

Figure 4.4 Temperature dependencies of (a) electrical resistivity and (b) thermoelectric power for $NaCo_2O_4$ [4].

to emphasize here is realization of a large S exhibiting a low ρ comparable to simple metals possessing S with one digit, which is indicative of a huge S for Na_xCoO_2.

4.4 Other Co Oxides

Extensive investigations following the report on Na_xCoO_2 resulted in the findings of other potential candidates for thermoelectric materials in Co oxides. For example, similar-layered Co oxides, $Ca_3Co_4O_9$ and $Bi_2Sr_3Co_2O_9$, show good thermoelectric performance [7, 8]. Structurally, they share common components, that is, the CoO_2 and block layers. The CoO_2 layer consists of a 2D triangular lattice of edge-sharing CoO_6 octahedra in the ab plane. In the block layer, on the other hand, cations and O^{2-} ions make a rock-salt lattice. Triple and quadruple subsystems form the block layer (see Fig. 4.5). The overall crystal structure of these materials consists of alternating layers of the CoO_2 and block layers stacked along the c axis.

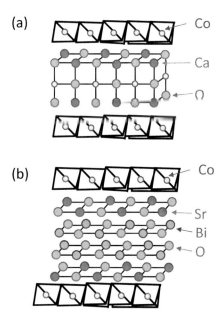

Figure 4.5 Crystal structures of (a) $Ca_3Co_4O_9$ and (b) $Bi_2Sr_3Co_2O_9$.

Apart from the 2D system, $La_{0.95}Sr_{0.05}CoO_3$ with the perovskite structure is claimed to be an efficient room-temperature thermoelectric oxide ($ZT = 0.18$) [9] The quasi-one-dimensional (Q1D) $Ca_3Co_2O_6$, which is a decomposed phase of $Ca_3Co_4O_9$, is shown to possess good thermoelectric efficiency at high temperature up to 1300 K [10]. Its ZT reaches 0.44 at 1300 K and tends to increase upon further heating. Interestingly, the temperatures where ZT takes a maximum value seem to display a trend toward an increase with decreasing dimensionality, that is, 300 K, 800 K, and above 1300 K for 3D, 2D, and 1D structures, respectively, among narrowing down these Co oxides. One can confirm how ZT behaves with temperature in Fig. 4.6.

4.5 Origin of a Huge Thermoelectric Power

The theory of strongly correlated electron systems, in which the degree of freedom of charge, spin, and orbital is taken into account,

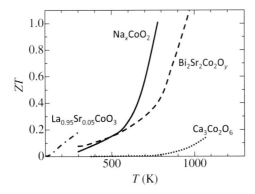

Figure 4.6 Temperature dependence of the dimensionless figure of merit for representative cobalt oxides [5, 9–11].

has shown that not only the larger degeneracies of Co ions but also their ratio together with the valence of Co ions are important for an enhancement in S. This situation naturally leads to a question whether the theory describes uniquely S of the real system, in which the valence and the spin state are changed systematically.

LaCoO$_3$ is an already well-studied system with a rather simple crystal structure and is considered to be a charge transfer insulator. Doping with Sr is easy, which alters the system to a metal *via* a metal–insulator transition, and then the electronic specific heat coefficient is much larger than simple metals, say, about 45 mJ/molK^2 when 20% of La is substituted by Sr.

Herefrom, the relationship between the electronic specific heat and the electron correlation is digested. The internal energy of the conduction electron system is expressed as

$$U_{el} = \int_0^\infty E\,N(E)\,f(E)\,dE. \tag{4.10}$$

This equation is the integral including the Fermi–Dirac function, so Eq. 1.61 is applicable, which can be imagined easily by replacing $\int_0^E E\,N(E)\,dE$ with $F(E)$.

Thus, U_{el} is far expanded as

$$U_{el} = \int_0^{E_{F(T)}} E\,N(E)\,dE + \frac{\pi^2}{6}(k_B T)^2 \left(\frac{d(E\,N(E))}{dE}\right)_{E=E_F(T)} + \cdots. \tag{4.11}$$

The first term of the right becomes

$$\int_0^{E_{F(T)}} E\,N(E)\mathrm{d}E = \int_0^{E_{F(0)}} E\,N(E)\mathrm{d}E + \int_{E_{F(0)}}^{E_{F(T)}} E\,N(E)\mathrm{d}E \quad (4.12)$$

$$= U_0 + [F(E)]_{E_F(0)}^{E_F(T)} \quad (4.13)$$

$$= U_0 + F(E_F(T)) - F(E_F(0)) \quad (4.14)$$

$$= U_0 + F(E_F(0)) + (E_F(T) - E_F(0))$$
$$\times \left(\frac{\mathrm{d}F(E)}{\mathrm{d}E}\right)_{E=E_F(0)} - F(E_F(0)) \quad (4.15)$$

$$= U_0 + F(E_F(0)) + (E_F(T) - E_F(0))$$
$$\times E_F(0)N(E_F(0)) - F(E_F(0)). \quad (4.16)$$

The second term of the right, on the other hand, is rewritten by the partial as

$$\frac{\pi^2}{6}(k_B T)^2 N(E_F(0)) + \frac{\pi^2}{6}(k_B T)^2 E_F(0)\left(\frac{\mathrm{d}(N(E))}{\mathrm{d}E}\right)_{E=E_F(T)}. \quad (4.17)$$

Thus,

$$U_{\mathrm{el}} = U_0 + (E_F(T) - E_F(0))E_F(0)N(E_F(0)) + \frac{\pi^2}{6}(k_B T)^2 N(E_F(0))$$
$$+ \frac{\pi^2}{6}(k_B T)^2 E_F(0)\left(\frac{\mathrm{d}(N(E))}{\mathrm{d}E}\right)_{E=E_F(T)} + E\cdots. \quad (4.18)$$

The formula of

$$E_F(T) = E_F(0) - \frac{\pi^2}{6}(k_B T)^2\left(\frac{(\mathrm{d}N(E)/\mathrm{d}E)_{E=E_F(T)}}{N(E_F(0))}\right) \quad (4.19)$$

can be applied for the second term. Finally, we obtain

$$U_{\mathrm{el}} = U_0 - \frac{\pi^2}{6}(k_B T)^2 E_F(0)\left(\frac{\mathrm{d}(N(E))}{\mathrm{d}E}\right)_{E=E_F(T)}$$
$$+ \frac{\pi^2}{6}(k_B T)^2 E_F(0)\left(\frac{\mathrm{d}(N(E))}{\mathrm{d}E}\right)_{E=E_F(T)}$$
$$+ \frac{\pi^2}{6}(k_B T)^2 N E_F(0) + E\cdots \quad (4.20)$$

$$= U_0 + \frac{\pi^2}{6}(k_B T)^2 N(E_F(0)) + \cdots. \quad (4.21)$$

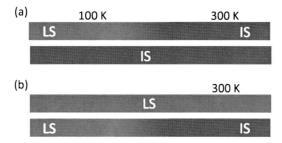

Figure 4.7 Spin states of Co^{3+} for (a) $R = La$ and (b) $R = Pr$, Nd, and Sm in $R_{1-x}Sr_xCoO_3$. The upper bar: $x = 0$ and the lower bar: $x \neq 0$.

Here U_0 is the internal energy of the conduction electron system at absolute zero temperature. The electronic specific heat is derived by dividing U by temperature as follows:

$$C_{el} = \frac{\partial U_{el}}{\partial T} = \frac{\pi^2}{3} k_B^2 N(E_F(0)) T. \tag{4.22}$$

The electron density of states for a free-electron model is yielded by

$$N(E) = \frac{dN}{dE} = \left(\frac{V}{2\pi^2}\right) \left(\frac{2m}{\hbar}\right)^{3/2} \sqrt{E}. \tag{4.23}$$

Thus, $C_{el} \propto m^{3/2}$; that is, a strong correlation effect is reflected as the effective mass. It can be concluded, from what has been mentioned above, that $La_{1-x}Sr_xCoO_3$ is a good example to study the carrier dependence of thermoelectric power in a strongly interacting electron system.

It is also possible to substitute La by another rare-earth (R) element as well, and we carried out a systematic study of the thermoelectric properties of $R_{1-x}Sr_xCoO_3$ with $R = La$, Pr, Nd, and Sm [12, 13]. To compare our results with the theory, we first need to digest the spin state of Co ions. We can represent the spin states in a simple diagram as Fig. 4.7. Sr-undoped $LaCoO_3$ undergoes a spin-state transition at about 100 K from the LS state to higher-spin states as temperature is increased. The spin state at high temperature has been a matter of argument for decades, although the IS state seems to be prevailing. For Sr-doped $LaCoO_3$, magnetic susceptibility measurements at high temperature suggest that Co^{3+} and Co^{4+} ions are both in the IS state at room temperature [14].

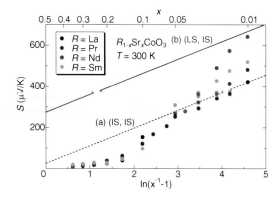

Figure 4.8 Thermoelectric power at 300 K for $R_{1-x}Sr_xCoO_3$ (R = La, Pr, Nd, and Sm) plotted as a function of $\ln(x^{-1}-1)$, where x is the Sr content shown on the upper axis [13]. The lines indicate the expectation of the theory by Koshibae et al. when (a) Co^{3+} and Co^{4+} ions have both the IS configuration and (b) Co^{3+} and Co^{4+} are, respectively, in the LS and IS states. *Abbreviations*: LS, low spin; IS, intermediate spin.

An X-ray absorption spectroscopy measurement combined with the cluster calculation has also concluded that Co^{4+} of $La_{1-x}Sr_xCoO_3$ is in the IS state [15]. Therefore, the theoretical curve for the case of both ions being in the IS state is drawn as line (a) in Fig. 4.7. On the other hand, Co^{3+} is in the LS state in the Sr-nondoped samples with R = Pr, Nd, and Sm. For LS Co^{3+} and IS Co^{4+}, the theory predicts line (b).

In the case of R = La, the experimental results followed fairly well line (a) up to $x = 0.05$, which is described by inserting $g_3 = 18$ and $g_4 = 24$ into Eq. 1.77. On the other hand, the Sr content dependence of S was quite different from the theoretical curve (b) when R = Pr, Nd, and Sm. Note that the slope of the theoretical curve is the same whatever the spin configuration is, as is obvious from Eq. 1.77, if the spin states do not depend on x. The faster decrease in S with increasing x compared to the theoretical line (b) for R = Pr, Nd, and Sm may be attributed to the increased population of the higher spin states of Co^{3+}. The high temperature limit for which Eq. 1.77 was derived corresponds to $t \ll k_B T$, where t denotes the transfer integral. This condition is not fulfilled when t is large, and it is necessary to calculate S at finite temperature. The deviation of the experimental data from curve (b) beyond $x = 0.05$ is probably because t already started to increase, violating the assumption of the theory. Although S decreases with increasing x, the electrical resistivity drops more rapidly, especially in the region of $0 < x < 0.1$ [13, 16]. Therefore, the power factor S^2/ρ takes a maximum at $x = 0.05$–0.1, and its value around room temperature is comparable to that of polycrystalline $NaCo_2O_4$ [17].

We can also apply Eq. 1.77 to the thermoelectric power of $NaCo_2O_4$. Since the average valence of a cobalt ion in the stoichiometric compound is 3.5, we obtain $x = 0.5$ and then do the thermoelectric power of 154 μV/K at a high temperature limit. Here we assume the relationship $g_3/g_4 = 1/6$ because magnetic data suggests both Co^{3+} and Co^{4+} are in the LS state. The experimental prediction seems to agree well with the experimental value, 150 μV/K. As shown above, the theory by Koshibae et al. yielded a large influence for understanding the essence of a huge S observed for Co oxides, especially from the viewpoint of the internal degree of freedom of electrons. Although the validity of the theory has been argued excitedly when we apply it to the real system, the most attractive point may be that one can easily speculate the value and sign of S on the basis of information about the valence and degeneracy of the transition metal without specific experiments.

Ong's group gave the crucial evidence for the spin entropy term by the complete suppression of thermoelectric power in a longitudinal magnetic field [18]: the possibility function is written as

$$Z = e^{\epsilon/k_B T} + e^{-\epsilon/k_B T}, \qquad (4.24)$$

where ϵ represents the energy. Then, Helmholtz energy is

$$F = -k_B T \log Z = -k_B T \log(e^{\epsilon/k_B T} + e^{-\epsilon/k_B T}). \qquad (4.25)$$

Using this expression, the entropy is derived:

$$s = \frac{\partial F}{\partial T} = k_B \log[2\cosh(\epsilon/k_B T)] - \frac{\epsilon}{T}\tanh(\epsilon/k_B T). \qquad (4.26)$$

Here, $\epsilon = g\mu_B H/2k_B T$ under the magnetic field H, and g, μ_B, k_B are the g factor, the Bohr magneton, and the Boltzmann constant, respectively. The thermoelectric power plotted against H/T provided a close fit by Eq. 4.26.

In contrast to the above localized picture, the metallic conductivity reminds us of the itinerant picture, in which extended Bloch states are formed. The transport properties are generally known to be determined by the electronic structure near the Fermi level, which has motivated research on its structure from the experimental and theoretical sides. The thermoelectric power was calculated using the Boltzmann transport equation with the electronic structure obtained by photoemission spectroscopy (PES) [19]. With Eq. 1.59 in mind, the thermoelectric power is expressed as

$$S = \frac{1}{eT} \frac{\int \sigma(\epsilon)(\epsilon - \zeta)\frac{\partial f_0}{\partial \epsilon} d\epsilon}{\int \sigma(\epsilon)\frac{\partial f_0}{\partial \epsilon} d\epsilon}. \tag{4.27}$$

Here, the spectral conductivity obtained from the Boltzmann equation is given by

$$\sigma(\epsilon) = \frac{e^2}{3} N(\epsilon) v^2(\epsilon) \tau(\epsilon, T). \tag{4.28}$$

In our analysis, we deduced information about $N(\epsilon)$ by performing PES measurements because the spectral intensity is proportional to the density of states. Furthermore, the above equation is rewritten by assuming the isotropic electronic structure in the ab plane, that is, $v(\epsilon) = \sqrt{2(\epsilon_{max} - \epsilon)/m_x}$ and an energy-independent relation time τ in a narrow energy range where $\partial f/\partial \epsilon$ takes a finite value as

$$S = \frac{1}{eT} \frac{\int |\epsilon_{max} - \epsilon| N(\epsilon)(\epsilon - \zeta)\frac{\partial f_0}{\partial \epsilon} d\epsilon}{\int |\epsilon_{max} - \epsilon| N(\epsilon)\frac{\partial f_0}{\partial \epsilon} d\epsilon}, \tag{4.29}$$

where ϵ_{max} is the highest energy. As a result, the calculated thermoelectric power is roughly consistent with the measured one. However, its value and temperature dependence seem to deviate notably for the systems with relatively high electrical resistivity. As for the thermoelectric power of metallic $Na_x CoO_2$, Singh evaluated the thermoelectric power to be 110 μV/K at 300 K on the basis of the band calculation [20], while Kuroki and Arita pointed out that the valence band of a peculiar shape called the pudding mold band made of d_{xy} orbitals is attributable to a large thermoelectric power and low electrical resistivity [21]. The upper end of the pudding mold band is relatively flat around Γ point, whereas the lower end has dispersion, some extended far from the Γ point. In this situation, the group velocity defined as $\nabla\epsilon(\mathbf{k})/\hbar$ is quite different above and below E_F locates the turning point in the band, resulting in a large thermoelectric power. In addition to the group velocity, a large Fermi surface makes the resistivity low.

Interestingly, some experiments imply the coexistence of electrons with the itinerant and localized nature. To take one example from some, the separation between two phases, the spin density wave and ferromagnetic phases, was found in the CoO_2 layer consisting of a crystallographically unique Co site by nuclear magnetic resonance (NMR) measurements and the degree of competition between them depended on the oxygen contents [22]. Combining the results of PES experiments, their phases are, respectively, expected to lead the paramagnetic metallic state and charge-ordered insulating state [23]. Furthermore, the volume fraction of these phases is found to change systematically depending on temperature and carrier concentration controlled by the oxygen amount.

4.6 Application

Last but not least are the following four conditions that should be overcome to achieve a wide variation of thermoelectric technology in the world. Specially, (i) device efficiency above 15%, (ii) power density above 1 W/cm^2, (iii) rich and environmentally friendly elements, and (iv) low cost 1 $/W are required. In particular, condition (i) is the most important issue. If we go beyond the barrier of 10% as system efficiency, a new market will be acquired. To realize this goal, we need $ZT \geq 2$ at least. Quite recently, Biswas et al. demonstrated ZT of 2.2 at 915 K in PbTe endotaxially nanostructured with 4 mol% SrTe [24]. The first commercial thermoelectric generator was oil-burning lamp powering made of ZnSb and constantan in 1948 [25]. Even at the present state, the thermoelectric technology is actively involved as a thermoelectric watch, the power source of a space exploration ship, a wine cooler, a cooling device for personal computers, and a gas sensor. Although the heat source depends on the purpose for which the device is used, in the case of thermoelectric generators for space, the electric power is produced from a Pu238 heat source, which is a radioisotope thermoelectric generator. This technology has been widely used in Apollo, Pioneer, Viking, and Voyager by NASA. Also, three batteries known as a solar cell, a fuel cell, and a storage cell have a high name recognition rate. In the present circumstances, it means that we map out a strategy for a thermoelectric generator as a fourth cell.

Exercises

4.1 The thermoelectric power at a high temperature limit is expressed by Eq. 1.77. First find the degeneracy of Co ions for each spin state. Then, describe the thermoelectric power as a function of the average valence of a Co ion under several spin-state combinations.

4.2 In the case of application, the *n*-type thermoelectric materials are also required. Show which oxide material is considered to be suitable for overcoming this issue on the basis of the valence and degeneracy of transition metal ions.

References

1. Goldsmid HJ (1958), The electrical conductivity and thermoelectric power of bismuth telluride, *Proc. Phys. Soc. Lond.*, **71**, 633–646.

2. Mahan G, Sales B, Sharp J (1997), Thermoelectric materials: new approaches to an old problem, *Phys. Today*, **50**, 42–47.

3. Tanaka T, Nakamura S, Iida S (1994), Observation of distinct metallic conductivity in $NaCo_2O_4$, *Jpn. J. Appl. Phys.*, **33**, L581–L582.

4. Terasaki I, Sasago Y, Uchinokura K (1997), Large thermoelectric power in $NaCo_2O_4$ single crystals, *Phys. Rev. B*, **56**, R12685–R12687.

5. Fujita K, Mochida T, Nakamura K (2001), High-temperature thermoelectric properties of $Na_xCoO_{2-\delta}$ single crystals, *Jpn. J. Appl. Phys.*, **40**, 4644–4647.

6. Jansen VM, Hopper R (1974), Notiz zur Kenntnis der Oxocobaltate des Natriums, *Z. Anorg. Allg. Chem.*, **408**, 104–106.

7. Masset AC, Michel C, Maignan A, Hervieu M, Toulemonde O, Studer F, Raveau B, Hejtmanek J (2000), Misfit-layered cobaltite with an anisotropic giant magnetoresistance: $Ca_3Co_4O_9$, *Phys. Rev. B*, **62**, 166–175.

8. Miyazaki Y, Miura T, Ono Y, Kajitani T (2002), Preparation and low-temperature thermoelectric properties of the composite crystal $[Ca_2(Co_{0.65}Cu_{0.35})_2O_4]_{0.624}CoO_2$, *Jpn. J. Appl. Phys.*, **41**, L849–L851.

9. Androulakis J, Migiakis P, Giapintzakis J (2004), $La_{0.95}Sr_{0.05}CoO_3$: an efficient room-temperature thermoelectric oxide, *Appl. Phys. Lett.*, **84**, 1099–1101.

10. Mikami M, Funahashi R, Yoshimura M, Mori Y, Sasaki T (2003), High-temperature thermoelectric properties of single-crystal $Ca_3Co_2O_6$, *J. Appl. Phys.*, **94**, 6579–6582.

11. Funahashi R, Shikano M (2002), $Bi_2Sr_2Co_2O_y$ whiskers with high thermoelectric figure of merit, *Appl. Phys. Lett.*, **81**, 1459–1461.

12. Takami T, Zhou JS, Goodenough JB, Ikuta H (2007), Correlation between the structure and the spin state in $R_{1-x}Sr_xCoO_3$ ($R =$ La, Pr, and Nd), *Phys. Rev. B*, **76**, 144116-1–144116-7.

13. Takami T, Ikuta H, Mizutani U (2004), Thermoelectric properties of $(R,Sr)CoO_3$ ($R =$ La, Pr, Nd, Sm), *Trans. Mater. Res. Soc. Jpn.*, **29**, 2777–2780.

14. Wu J, Leighton C (2003), Glassy ferromagnetism and magnetic phase separation in $La_{1-x}Sr_xCoO_3$, *Phys. Rev. B*, **67**, 174408-1–174408-16.

15. Potze RH, Sawatzky GA, Abbate M (1995), Possibility for an intermediate-spin ground state in the charge-transfer material $SrCoO_3$, *Phys. Rev. B*, **51**, 11501–11506.

16. Fujishiro H , Fujine Y, Kashiwada Y, Ikebe M, Hejtmanek J (2003), Enhanced thermoelectric properties at x 0.1 in $La_{1-x}Sr_xCoO_3$ and $La_{1-x}Sr_x(Co_{1-y}M_y)O_3$ ($M =$ Cr, Cu), *Proc. 22nd Int. Conf. Thermoelectr.*, 235–238.

17. Takahata K, Iguchi Y, Tanaka D, Itoh T, Terasaki I (2000), Low thermal conductivity of the layered oxide $(Na,Ca)Co_2O_4$: another example of a phonon glass and an electron crystal, *Phys. Rev. B*, **61**, 12551–12555.

18. Wang Y, Rogado NS, Cava RJ, Ong NP (2003), Spin entropy as the likely source of enhanced thermopower in Na_xCoO_2, *Nature*, **423**, 425–428.

19. Takeuchi T, Kondo T, Takami T, Takahashi H, Ikuta H, Mizutani U, Soda K, Funahashi R, Shikano M, Mikami M, Tsuda S, Yokoya T, Shin S (2004), Contribution of electronic structure to the large thermoelectric power in layered cobalt oxides, *Phys. Rev. B*, **69**, 125410-1–125410-9.

20. Singh DJ (2000), Electronic structure of $NaCo_2O_4$, *Phys. Rev. B*, **61**, 13397–13402.

21. Kuroki K, Arita R (2007), Pudding moldh band drives large thermopower in Na_xCoO_2, *J. Phys. Soc. Jpn.*, **76**, 083707-1–083707-4.

22. Takami T, Nanba H, Umeshima Y, Itoh M, Nozaki H, Itahara H, Sugiyama J (2010), Phase separation in the CoO_2 layer observed in thermoelectric layered cobalt dioxides, *Phys. Rev. B*, **81**, 014401-1–014401-12.

23. Wakisaka Y, Hirata S, Mizokawa T, Suzuki Y, Miyazaki Y, Kajitani T (2008), Electronic structure of $Ca_3Co_4O_9$ studied by photoemission spectroscopy: phase separation and charge localization, *Phys. Rev. B*, **78**, 235107-1–235107-6.

24. Biswas K, He J, Blum ID, Wu CI, Hogan TP, Seidman DN, Dravid VP, Kanatzidis MG (2012), High-performance bulk thermoelectrics with all-scale hierarchical architectures, *Nature*, **489**, 414–418.

25. Vedernikov MV, Iordanishvili EK (1998), A. F. Ioffe and origin of modern semiconductor thermoelectric conversion, *Proc. of 17th Int. Conf. Thermoelectr.*, **1**, 37–42.

Chapter 5

Room-Temperature Ferromagnetism

5.1 Prologue

Ferromagnetism is the fundamental property by which certain materials form permanent magnets or are attracted to magnets. The representative ferromagnetic (FM) materials are Fe, Co, Ni, most of their alloys, some compounds of rare-earth metals, and a few naturally occurring minerals like lodestone. In addition to the basic side, ferromagnetism is very important in industry and modern technology and is the basis for many electrical and electromechanical devices such as electric motors, transformers, magnetic storage like tape recorders, and hard disks.

One of the fundamental properties of an electron is that it has a dipole moment, that is, it behaves itself as a tiny magnet. This dipole moment comes from the more fundamental property of the electron, namely, it has quantum mechanical spin. The quantum mechanical nature of this spin causes the electron to be able to be restricted in only two states with the magnetic field either pointing up or down. The spin of electrons in atoms is the main source of the magnetism not just limited to ferromagnetism, although there is also a contribution from the orbital angular momentum of the electron around the nucleus.

Functional Cobalt Oxides: Fundamentals, Properties, and Applications
Tsuyoshi Takami
Copyright © 2014 Pan Stanford Publishing Pte. Ltd.
ISBN 978-981-4463-32-4 (Hardcover), 978-981-4463-33-1 (eBook)
www.panstanford.com

Table 5.1 FM metals and Curie temperature

Material	Curie temperature
Fe	1043 K
Co	1388 K
Ni	627 K
Gd	292 K

For materials with a filled electron shell, the total dipole moment of the electrons is zero because the spins are in up/down pairs. Only atoms with a partially filled shell, on the contrary, can have a net magnetic moment; thus ferromagnetism occurs only for materials under this condition. This is the quantum model in which whether magnetic or nonmagnetic is determined by whether electrons with spins on the orbital are paired or unpaired.

Table 5.1 shows representative FM metals; Tb, Dy, Ho, Er, and Tm are also FM materials as an element. When the total number of electrons with spin-up and spin-down are written as $N_\uparrow$ and $N_\downarrow$, the total number of electrons N and the quantity proportional to the magnetization M are $N = N_\uparrow + N_\downarrow$ and $M = N_\downarrow - N_\uparrow$, respectively. The exchange energy is given by

$$\epsilon_{ex} = J\, N_\uparrow \cdot N_\downarrow = \frac{1}{4}J\,(N^2 - M^2), \qquad (5.1)$$

using the Coulomb interaction J. The magnetization originates from relative shift of the spin band. An increase in the kinetic energy is expressed as

$$\Delta\epsilon_{kin} = \Delta\left[N\left(\epsilon_F + \frac{\Delta}{2}\right)\cdot\left(\epsilon_F + \frac{\Delta}{2}\right) - N\left(\epsilon_F - \frac{\Delta}{2}\right)\cdot\left(\epsilon_F - \frac{\Delta}{2}\right)\right]$$
$$(5.2)$$

$$\approx \Delta^2\left[N(\epsilon_F) + \epsilon_F\left(\frac{\partial N(\epsilon)}{\partial\epsilon}\right)_{\epsilon=\epsilon_F}\right] = \Delta^2 N(\epsilon_F), \qquad (5.3)$$

given the magnitude of the shift 2Δ. The above equation is derived after the Taylor expansion:

$$N\left(\epsilon_F \pm \frac{\Delta}{2}\right) = N(\epsilon_F) \pm \frac{\Delta}{2}\left(\frac{\partial N(\epsilon)}{\partial\epsilon}\right)_{\epsilon_F} + \cdots. \qquad (5.4)$$

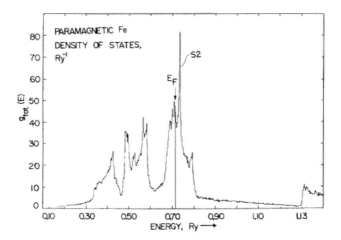

Figure 5.1 Density of states of Fe [1]. E_F and S2 denote the Fermi energy and the singularity, respectively.

On the other hand, the magnetization is written as

$$M = \int_0^{\epsilon_F} N(\epsilon)d\epsilon - \int_{2\Delta}^{\epsilon_F} N(\epsilon - 2\Delta)d\epsilon \tag{5.5}$$

$$= \int_0^{\epsilon_F} N(\epsilon)d\epsilon - \int_0^{\epsilon_F - 2\Delta} N(z)dz \tag{5.6}$$

$$= \int_0^{\epsilon_F} N(\epsilon)d\epsilon - \int_0^{\epsilon_F} N(\epsilon)d\epsilon - \int_{\epsilon_F}^{\epsilon_F - 2\Delta} N(\epsilon)d\epsilon \tag{5.7}$$

$$\approx 2\Delta N(\epsilon_F). \tag{5.8}$$

Thus,

$$\Delta\epsilon_{kin} = M^2/4N(\epsilon_F). \tag{5.9}$$

Finally, the total energy becomes

$$\epsilon = \epsilon_{ex} + \Delta\epsilon_{kin} = \frac{1}{4}\left(\frac{1}{N(\epsilon_F)} - J\right)M^2 + \cdots . \tag{5.10}$$

The condition of the negative coefficient leads to $J N(\epsilon_F) > 1$, which means that a large $N(\epsilon_F)$ as well as J is required. For example, this situation is fulfilled for Fe (see Fig. 5.1).

There is another insight for understanding magnetism. Given all the atoms have magnetic spins independently and the magnetic

interactions between them exists, the difference between paramagnetism, ferromagnetism, and antiferromagnetism is excellently
interpreted. In the case of ferromagnetism of oxide materials,
spins are aligned along the external field by the double-exchange
interaction. In the following sections, ferromagnetism in transition
metal oxides is introduced. Among them, the magnetic properties
of $(Sr,Y)CoO_{3-\delta}$ are explained as an example that shows the highest
FM transition temperature (T_c) of approximately room temperature
among cobalt oxides.

5.2 Ferromagnetism in Transition Metal Oxides

As for the history of ferromagnetism, P. Curie discovered that
ferromagnetism disappeared above a certain temperature and the
magnetization is reciprocal to temperature [2]. P. Langevin regarded
the atomic group with a magnetic moment of a constant magnitude
and applied classical statistical mechanics, resulting in derivation
of the Curie law under the magnetic field [3]. P. Weiss introduced
the interaction that acts to align the magnetic moments parallel
each other [4]. However, this theory could not explain the coefficient
Γ defined as $\mathbf{H}/\mathbf{M}$ and the classical physical description, in which
the assumption of atomic magnetic moment does not obtain a
consensus with the justification that the magnetism originates
from the current, where $\mathbf{H}$ and $\mathbf{M}$ are the internal fields and the
magnetization, respectively.

W. Heisenberg proposed the theory of ferromagnetism on the
basis of exchange interaction between electron spins in atoms and
overcame the issue by the well-known Heisenberg model [5]. F.
Bloch developed the theory of ferromagnetism originating from the
solid-state electron theory [6]. He applied the electron-gas model
under the mean-field approximation and obtained the result that the
ferromagnetism appears below a certain density.

Transition metal oxides with a perovskite structure exhibit
many interesting phenomena such as superconductivity, colossal
magnetoresistance, and ferroelectricity. If the topic is limited to
ferromagnetism herein, among perovskite Mn oxides, the maximum
T_c to the FM phase is 370 K observed for $La_{0.7}Sr_{0.3}MnO_3$ [7]. The

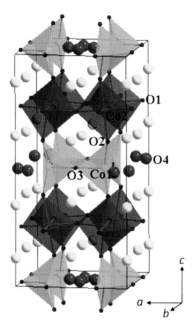

Figure 5.2 Crystal structure of $Sr_{0.7}Y_{0.3}CoO_{2.63}$. The Co atom locates in the polyhedral, and the yellow symbol represents Sr or Y [11].

ferromagnetism has been accounted for by the double-exchange interaction between transition metal ions. Among Co oxides, for instance, $(La,Sr)CoO_3$ and $RBaCo_2O_{5.5}$ have been known to show ferromagnetism and their T_c are 270 K and $\approx$ 300 K, respectively [8, 9]. However, the Co oxide that undergoes an FM transition above room temperature was still lacking. In recent years, it was found that $(Sr,Y)CoO_{3-\delta}$ can be regarded as a room-temperature ferromagnet [10]; its crystal structure and magnetic properties are shown in the next section.

5.3 Room-Temperature Ferromagnetism of (Sr,Y)CoO$_{3-\delta}$

Where $AMO_{3-\delta}$ (A: the alkaline-earth metal, M: the transition metal) oxygen-deficient perovskites contain an M cation that is stable in a tetrahedral as well as an octahedral site, the oxygen vacancies order into $MO_{2(1-\delta)}$ (001) planes that alternate with

MO_2 (001) planes of corner-shared $MO_{6/2}$ octahedral sites. In the brownmillerite structure of $Ca_2Fe_2O_5$, the order of oxygen vacancies further increases within oxygen-deficient planes to create corner-sharing tetrahedral sites [12]. In the range $0.5 < \delta < 0.75$, ordering of the oxygen vacancies within the oxygen-deficient layers may accommodate M atoms in corner-sharing tetrahedral and square-pyramidal sites. The crystal structure of the as-grown sample of $Sr_{0.7}Y_{0.3}CoO_{2.63}$ is illustrated in Fig. 5.2 [11]. Sr, Y, and Co are ordered, whereas the O atom in the O4 site is deficient [12]. The ordered ab planes are stacked along the c axis, resulting in a quadruple of the c axis length compared to the primitive cell.

Figure 5.3a shows the magnetization as a function of temperature for $Sr_{1-x}Y_xCoO_{3-\delta}$. The samples with $0.2 \leq x \leq 0.25$ show a large magnetization, and T_c is estimated to be approximately 335 K, not depending particularly on x, which is the highest temperature among perovskite cobalt oxides, whereas those with $x < 0.2$ and >0.25 do not exhibit a clear FM behavior. To investigate the oxygen effect on ferromagnetism, the magnetization of $Sr_{0.775}Y_{0.225}CoO_3$ $(\delta = 0)$ annealed in high-pressure oxygen is measured, which is shown in the inset of Fig. 5.3a. T_c is found to be remarkably decreased despite the large magnetization. We are now in a position to state that oxygen deficiency is essential to the observed room-temperature ferromagnetism.

Figure 5.3b shows the magnetic field dependence of the magnetization for $Sr_{0.775}Y_{0.225}CoO_{3-\delta}$ at 10 K. The maximum magnetization reaches 0.25 μ_B/Co in the range of fields up to 7 T, together with an apparent spontaneous magnetization, indicative of bulk ferromagnetism. Also, the $M(H)$ shape for $\delta = 0$ is different from that of the nonannealed samples, as well as the electrical resistivity data. These experimental facts further consolidate that the ferromagnetism for $x = 0.225$ is not caused by the double-exchange interaction that works between transition metal ions. The occurrence of the ferromagnetism in the narrow range of $0.2 \leq x \leq 0.25$ also supports the scenario that the double-exchange mechanism does not play a significant role and brings a remarkable contrast with $La_{1-x}Sr_xCoO_3$ [13]. Besides the oxygen site, the substitutions of 25% Co and 50% Co in $Sr_{0.75}Y_{0.25}CoO_{2.62+\delta}$ by Gd and Fe lower T_c down to 225 K and vanish T_c, respectively, regardless of an increase in oxygen content [14].

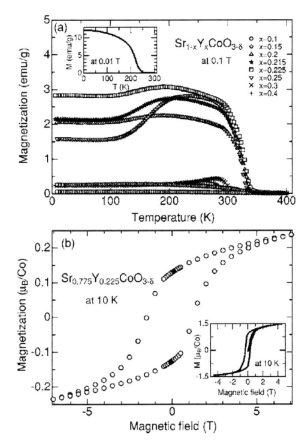

Figure 5.3 Magnetization is plotted as a function of (a) temperature for Sr$_{1-x}$Y$_x$CoO$_{3-\delta}$ ($x = 0.1$–0.4) and (b) magnetic field for Sr$_{0.775}$Y$_{0.225}$CoO$_{3-\delta}$. The insets in (a) and (b) are the data of Sr$_{0.775}$Y$_{0.225}$CoO$_3$ [10].

Next, the (103) peak in the X-ray diffraction patterns is remarked since it gives information on the Sr/Y site ordering. It appears above $x = 0.2$, which exactly corresponds to the emergence of the room-temperature ferromagnetism but still continues to exist beyond $x = 0.25$. At first sight, this may seem to be inconsistent with the fact that the $x = 0.3$ sample does not show bulk ferromagnetism. Because Y cannot occupy the Sr site for $x > 0.25$, the ferromagnetism may be quite susceptible to the Sr/Y solid solution in the Sr site, and thereby it may be concluded that the FM order arises from the

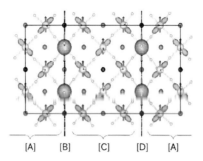

[A]   [B]   [C]   [D]   [A]

Figure 5.4 Orbital- and spin-ordering structure in the CoO_6 layer for $Sr_{0.75}Y_{0.25}CoO_{2.625}$ [15]. In regions (A) and (C), spins for the orbital ordering intermediate-spin state are coupled ferromagnetically, which form an antiferromagnitic (AF) order with the high-spin Co ions in regions (B) and (D).

orbital ordering of the intermediate-spin (IS) Co^{3+} ions, which is stabilized by the peculiar Y/Sr ordering.

The magnetic layers can be broken down two components, that is, the $CoO_{4.25}$ layer including the oxygen vacancies and the CoO_6 layer including the isotropic and anisotropic octahedra. AF ordering of the high-spin (HS) state is formed in the $CoO_{4.25}$ layer, but the local magnetism in the CoO_6 layer is more complex. The resonant X-ray scattering technique focused on near the Co K-edge energy gave crucial evidence of the presence of the IS state together with e_g orbital ordering [15]. As shown in Fig. 5.4, the orbital ordering for the IS state is formed between which FM superexchange interaction works in the regions (A) and (C) [16], whereas Co ions are in the HS state in the regions (B) and (D). An AF superexchange interaction is expected to work between the IS and HS states, resulting in an AF coupling between the regions (A)/(C) and (B)/(D).

References

1. Maglic R (1973), Van Hove singularity in the iron density of states, *Phys. Rev. Lett.*, **31**, 546–548.
2. Curie P (1894), Proprietes magnetiques du fer a diverses temperatures, *Compt. Rend.*, **118**, 796–800.

3. Langevin P (1905), Sur la theorie du magnetisme, *J. Phys. (Paris)*, **4**, 678–693.

4. Weiss P (1907), L'hypothese du champ moleculaire et la propriete ferromagnetique, *J. Phys. (Paris)*, **6**, 661–690.

5. Heisenberg W (1928), On the theory of ferromagnetism, *Z. Phys.*, **49**, 619–636.

6. Bloch F (1929), Note to the electron theory of ferromagnetism and electrical conductivity, *Z. Phys.*, **57**, 545–555.

7. Tokura Y (2000), *Colossal Magnetoresistive Oxides*, Gordon and Breach, New York.

8. Jonker GH, Van Santen JH (1953), Magnetic compounds with perovskite structure III. Ferromagnetic compounds of cobalt, *Physica*, **19**, 120–130.

9. Taskin AA, Lavrov AN, Ando Y (2002), Ising-like spin anisotropy and competing antiferromagnetic-ferromagnetic orders in $GdBaCo_2O_{5.5}$ single crystals, *Phys. Rev. Lett.*, **90**, 227201-1–227201-4.

10. Kobayashi W, Ishiwata S, Terasaki I, Takano M, Grigoraviciute I, Yamauchi H, Karppinen M (2005), Room-temperature ferromagnetism in $Sr_{1-x}Y_xCoO_{3-\delta}$ ($0.2 \leqslant x \leqslant 0.25$), *Phys. Rev. B*, **72**, 104408-1–104408-5.

11. Li Y, Kim YN, Cheng JG, Alonso JA, Hu Z, Chin YY, Takami T, Fernández-Díaz MT, Lin HJ, Chen CT, Tjeng LH, Manthiram A, Goodenough JB (2011), Oxygen-deficient perovskite $Sr_{0.7}Y_{0.3}CoO_{2.65-\delta}$ as a cathode for intermediate-temperature solid oxide fuel cells, *Chem. Mater.*, **23**, 5037–5044.

12. Istomin SY, Grins J, Svensson G, Drozhzhin OA, Kozhevnikov VL, Antipov EV, Attfield JP (2003), Crystal structure of the novel complex cobalt oxide $Sr_{0.7}Y_{0.3}CoO_{2.62}$, *Chem. Mater.*, **15**, 4012–4020.

13. Señarís-Rodríguez MA, Goodenough JB (1995), Magnetic and transport properties of the system $La_{1-x}Sr_xCoO_{3-\delta}$ ($0 < x \leq 0.50$), *J. Solid State Chem.*, **118**, 323–336.

14. Lindberg F, Drozhzhin OA, Istomin SY, Svensson G, Kaynak FB, Svedlindh P, Warnicke P, Wannberg A, Mellergaård A, Antipov EV (2006), Synthesis and characterization of $Sr_{0.75}Y_{0.25}Co_{1-x}M_xO_{2.625+\delta}$ ($M = Ga$, $0.125 \leqslant x \leqslant 0.500$ and $M = Fe$, $0.125 \leqslant x \leqslant 0.875$), *J. Solid State Chem.*, **179**, 1434–1444.

15. Nakao H, Murata T, Bizen D, Murakami Y, Ohoyama K, Yamada K, Ishiwata S, Kobayashi W, Terasaki I (2011), Orbital ordering of

intermediate-spin state of Co^{3+} in $Sr_3YCo_4O_{10.5}$, *J. Phys. Soc. Jpn.*, **80**, 023711-1–023711-4.

16. Medarde M, Dallera C, Grioni M, Voigt J, Pomjakushina APE, Conder K, Neisius T, Tjernberg O, Barilo SN (2006), Low-temperature spin-state transition in $LaCoO_3$ investigated using resonant x-ray absorption at the Co K edge, Phys. Rev. B, 73, 054424-1 054424-10.

Chapter 6

Partially Disordered Antiferromagnetic Transition

6.1 Prologue

The magnetic structure is, broadly speaking, determined by the following four factors, that is, the type of interaction, the dimensionality, the spin quantum number, and the coordination number or the crystal structure. Being different from simple ferromagnetism, antiferromagnetism, and ferrimagnetism, several magnetic interactions sometimes coexist in materials, resulting in a complex magnetic structure. In this chapter, we discuss the partially disordered antiferromagnetic (PDA) transition realized via the novel ordering process for $Ca_3Co_2O_6$, research on which has been superficial and reported for limited one-dimensional (1D) materials such as $CsCoCl_3$.

To begin with, let's testify that the magnetic transition does not occur for a pure 1D system. The Hamiltonian of a 1D system is written as

$$\mathcal{H} = -J \sum_{i=1}^{N} \sigma_i \sigma_{i+1}, \tag{6.1}$$

provided the number of atoms N is large enough for defining N as an even number for clarity and the Ising model ($\sigma_i = \pm 1$). Then, the

Functional Cobalt Oxides: Fundamentals, Properties, and Applications
Tsuyoshi Takami
Copyright © 2014 Pan Stanford Publishing Pte. Ltd.
ISBN 978-981-4463-32-4 (Hardcover), 978-981-4463-33-1 (eBook)
www.panstanford.com

partition function is given by

$$Z = \sum_{\sigma_1=\pm 1} \sum_{\sigma_2=\pm 1} \cdots \sum_{\sigma_N=\pm 1} \exp(\alpha \sum_{+,-} \sigma_i \sigma_{i+1}), \qquad (6.2)$$

where $\alpha = J/k_B T$. When $\sigma_i \sigma_{i+1} = +1$ and -1, the exponential term becomes $\exp(\alpha)$ and $\exp(-\alpha)$, respectively. Thus, the above equation is converted to

$$\exp(\alpha \sigma_i \sigma_{i+1}) = \cosh\alpha + \sigma_i \sigma_{i+1}\sinh\alpha \qquad (6.3)$$

to fulfill the condition. The partition function is rewritten as

$$Z = \sum_{\sigma_1=\pm 1} \sum_{\sigma_2=\pm 1} \cdots \sum_{\sigma_N=\pm 1} \Pi(\cosh\alpha + \sigma_i \sigma_{i+1}\sinh\alpha) \qquad (6.4)$$

$$= 2^N(\cosh^N \alpha + \sinh^N \alpha) \qquad (6.5)$$

$$\approx (2\cosh\alpha)^N. \qquad (6.6)$$

As for the expansion of the term including $\sinh\alpha$,

$$\sinh^N \alpha(\sigma_1\sigma_2)(\sigma_2\sigma_3)\cdots(\sigma_{N-1}\sigma_N)(\sigma_N\sigma_1) = \sinh^N \alpha \sigma_1 \sigma_2^2 \sigma_3^2 \cdots \sigma_N^2 \sigma_1 \qquad (6.7)$$

$$= \sinh^N \alpha, \qquad (6.8)$$

where $N+1 \equiv 1$. The magnetic specific heat is calculated using the internal energy U and the Gibbs free energy G as

$$C_M = \frac{dU}{dT} \qquad (6.9)$$

$$= \frac{d\partial(G/T)/\partial(1/T)}{dT} \qquad (6.10)$$

$$= \frac{NJ^2}{k_B T^2}\cosh^2 \frac{J}{k_B T}. \qquad (6.11)$$

On these grounds the $C_M(T)$ curve is continuous, indicating no magnetic transition. In addition to magnetism, research on the 1D electron system has embraced many different fields such as a quantum wire, carbon nanotube, and 1D stripe observed for cuprate superconductors. The magnetic properties of the 1D homologous series $A_{n+2}Co_{n+1}O_{3n+3}$ (A: the alkaline-earth metal, n: the integer) including $Ca_3Co_2O_6$ ($A = Ca$, $n = 1$) are also presented with varying magnetic field and temperature.

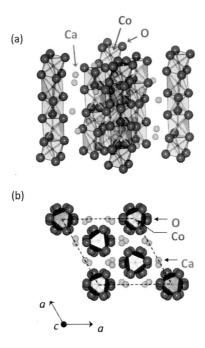

Figure 6.1 (a) Crystal structure and (b) *c*-plane view of Ca$_3$Co$_2$O$_6$.

6.2 Magnetism of Ca$_3$Co$_2$O$_6$

Ca$_3$Co$_2$O$_6$ is a compound that has been already known as 1D system for decades [1], and many studies have been devoted especially on its peculiar magnetic properties. Figure 6.1 shows the crystal structure of Ca$_3$Co$_2$O$_6$. The 1D chain consists of one CoO$_6$ trigonal prism and one CoO$_6$ octahedron, which are face-sharing alternatively and connected one-dimensionally along the *c* axis. One CoO$_3$ chain is surrounded by six nearest-neighbor CoO$_3$ chains in the two-dimensional (2D) triangular lattice in the *ab* plane (see Fig. 6.1b), which could produce geometrical frustration. Frustration is instability of the electron spin configuration, which takes place in the geometrical condition of the lattice when the antiferromagnetic (AF) interaction acts. To put it plainly, the situation where the magnetic ordering is not realized by the geometrical condition is called magnetic frustration and as its stage, for example, the

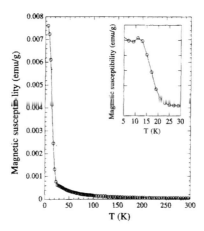

Figure 6.2 Temperature dependence of magnetic susceptibility for $Ca_3Co_2O_6$. Inset shows the magnetic susceptibility below 30 K as an expanded scale [2].

triangular lattice and the pyrochlore lattice. For instance, on the occasion of the spin distribution on the triangular lattice, the AF coupling is actualized between two spins, but the remaining one-third cannot be aligned either parallel or antiparallel; that is to say, the AF coupling is formed toward one spin, and yet the FM coupling toward another spin is forced at the same time. This is what frustration is all about. The way of connection for CoO_6 octahedra is in contrast with the crystal structure of $NaCo_2O_4$, $Ca_3Co_4O_9$, and $Bi_2Sr_3Co_2O_9$, which consists of edge-sharing CoO_6 octahedra forming a triangular lattice of Co, or with the crystal structure of $R_{1-x}Sr_xCoO_3$ (R: the rare-earth element), which consists of corner-sharing CoO_6 octahedra forming a three-dimensional (3D) network.

For $Ca_3Co_2O_6$, the magnetic transition to the PDA state occurs via a complex ordering process together with the complex temperature-field phase diagram due to the competition between the intrachain ferromagnetic (FM) and interchain AF interactions. Actually, a PDA state, in which two-thirds of the chains in the triangular lattice exhibit AF order, but the remaining one-third remains incoherent (disordered), is concluded to be realized for $Ca_3Co_2O_6$ at 20 K (T_N). As shown in Fig. 6.2, the remaining incoherent spins are aligned along the direction of the magnetic field, resulting in a marked

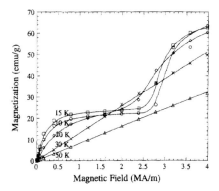

Figure 6.3 Magnetic field dependence of magnetization at 10 K, 15 K, 20 K, 30 K, and 50 K for $Ca_3Co_2O_6$ [2].

increase of the magnetic susceptibility below T_N [2]. The magnetic field variation of magnetization also displays a plateau with the magnitude of 1/3 of the full magnetic moment (see Fig. 6.3) [2], which is qualitatively consistent with two-thirds of the spins are parallel to the magnetic field, whereas one-third is antiparallel. Stronger magnetic fields make the antiparallel spins align toward the magnetic field, and the field-induced FM phase is formed, as can be confirmed from the data at 10 K and 15 K above $\approx$ 2.5 MA/m. With a further decrease of temperature, a spin-freezing behavior was also observed below 10 K, where the magnetic susceptibility measured under the field-cooled mode is almost independent of temperature. Above all, five stable magnetic configurations were evidenced by the magnetization versus magnetic field curve at 2 K. Their relative values at five steps are 1/4, 1/2, 1, 2, and 3, setting the full moment to 3. A. Maignan et al. proposed a possible model to explain a multistep $M(H)$ behavior (see Fig. 6.4) [3]. $Ca_3Co_2O_6$ nanoparticles, on the other hand, do not show multistep magnetization behavior widely reported for single crystals and polycrystalline samples despite the fact that the transition in $\chi(T)$ is essentially unaffected [4]. As another nanosized effect of $Ca_3Co_2O_6$, the thermal conductivity is fairly reduced due to the increased phonon scattering [5].

Now, let's flush back to the history and origin of the PDA state. For an Ising spin triangular lattice system with perfect frustration,

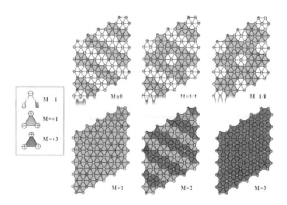

Figure 6.4 Possible models for magnetic configuration [3]. $+$ and $-$ show spins aligned parallel and antiparallel to the c axis, respectively.

spins are not ordered down to 0 K, as we have seen, in contrast to the square lattice system. However, a long-range magnetic order may be formed by the interaction apart from the nearest-neighbor interaction. Mekata solved theoretically the thermal equilibrium state of the system with the next nearest-neighbor interaction J' with a positive sign using the mean field theory at a finite temperature and pointed out the presence of the PDA state by varying the ratio of the nearest-neighbor interaction J and J' [6]. As shown in Fig. 6.5, he considered the magnetism by separating the whole lattice into sublattices characterized by three atomic sites (l, m, n). When J acts between the sublattices l and m, the number of the nearest magnetic atoms z' becomes 3. Taking into J between the sublattices l and n consideration, eventually $z' = 6$. In the case of J', on the other hand, a similar step brings about $z' = 12$. Therefore, the Hamiltonian of the sublattice specified as l can be written as

$$\mathcal{H}_l = -6J \left(< S_m > + < S_n > \right) \sum_i S_l^i - 12J' < S_l >$$

$$\sum_i S_l^i - g\mu_B H \sum_i S_l^i,$$

where i denotes the unit cell of the sublattice and S represents the spin quantum number. Hereafter the coefficients of $<S>\sum_i S_l^i$ and the third term are replaced by A and B, respectively. Then,

$$< S >= SB_s(x) \qquad (6.12)$$

and

$$x = (A < S > +B)S/k_B T \tag{6.13}$$

are obtained using the Brillouin function $B_s(x)$. The above relationship can be rewritten as

$$tx - h = B_s(x) \tag{6.14}$$

and

$$t = k_B T/AS < S >, h = B/A < S > . \tag{6.15}$$

When $x \to 0$, the Brillouin function is approximated as

$$B_s(x) = \frac{S+1}{3S}x - \frac{S+1}{3S}\frac{2S^2 + 2S + 1}{30S^2}x^3 + \cdots . \tag{6.16}$$

Thus, for $h = 0$, Eq. (6.15) is expanded as

$$\frac{k_B T_N}{6S^2 J \left(\frac{2J'}{J} < \sigma_l > + < \sigma_m > + < \sigma_n >\right)}x = \frac{S+1}{3S}x, \tag{6.17}$$

using the relative magnetization $< \sigma >$ defined as $S_l = S < \sigma_l >$. By applying the relationships of $<\sigma_l> + <\sigma_m> + <\sigma_n> = 0$ and $< \sigma_l > = 3S/(S+1)$,

$$T_N = \frac{6S(S+1)}{k_B}(2J' - J). \tag{6.18}$$

The magnetic phase diagram is shown in Fig. 6.6, in which α defined as $2J'/J$ is plotted as a function of normalized temperature. It follows from Fig. 6.6 that three magnetic phases are found to exist for $0 \leq |\alpha| \leq 0.8$.

However, interplane interactions are not actually ignored for the materials. The first example of a triangular AF material is CsCoCl$_3$, in which the quasi-1D AF chains form a triangular lattice in the *ab* plane [7]. As an interesting point, the novel temperature dependence of the magnetic reflection peak detected by neutron diffraction measurement is excellently analyzed by the above theory. For this compound, the AF interaction works in the 1D chain and the FM interaction does between 1D chains, which is contrary to the case of Ca$_3$Co$_2$O$_6$. It is no wonder that these ordered states and the geometrical frustration lead to a question about the magnetic ordering process from the static and dynamic sides. Recently, significant progress for understanding the process to the PDA state observed for Ca$_3$Co$_2$O$_6$ was obtained, which is explained in the next section.

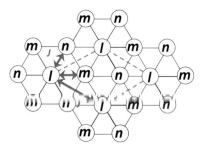

Figure 6.5 Sublattice formed in the *ab* plane of the triangular lattice system. The dashed line shows the unit cell of $Ca_3Co_2O_6$. J and J' represent the nearest and next nearest-neighbor interaction, respectively.

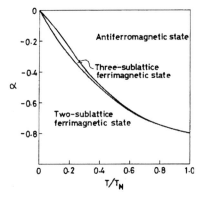

Figure 6.6 Ratio of J to $2J'$ is plotted as a function of temperature normalized by T_N as a magnetic phase diagram assuming the sublattice [6].

6.3 Magnetism of Quasi-1D $A_{n+2}Co_{n+1}O_{3n+3}$

$Ca_3Co_2O_6$ is the first member ($n = 1$) of the homologous series $A_{n+2}Co_{n+1}O_{3n+3}$ ($A =$ Ca, Sr, Ba), and more generally of the $A_{n+2}B'B_nO_{3n+3}$ ($B' =$ Ni, Cu, Zn; $B =$ Co, Ir, Pt) series with $A =$ Ca, $B' = B =$ Co, and the same may be said, no doubt, of their crystal structure. The unit cell of $A_{n+2}Co_{n+1}O_{3n+3}$ includes columnar structures, between which the alkaline-earth metal A is located [8]. The Co–O columnar structure consists of one CoO_6 trigonal prism and n CoO_6 octahedra, which are face-sharing and connected one-dimensionally, as illustrated in Fig. 6.7. For the other end member,

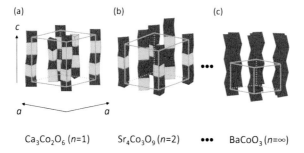

(a)   (b)   (c)

$Ca_3Co_2O_6$ ($n=1$) $Sr_4Co_3O_9$ ($n=2$)   ••• $BaCoO_3$ ($n=\infty$)

Figure 6.7 Crystal structures of (a) $n = 1$, (b) $n = 2$, and (c) $n = \infty$ in $A_{n+2}Co_{n+1}O_{3n+3}$. Red and blue polyhedra show the CoO_6 octahedron and the CoO_6 trigonal prism, respectively.

$BaCoO_3$ ($A = Ba$, $n = \infty$), crystallizes in the hexagonal perovskite structure, in which only the face-sharing CoO_6 octahedra form a 1D CoO_3 chain along the c axis [9].

The ionic radius of A ions need to be increased with n from Ca to Sr and to Ba to obtain the single-phased samples, which is probably due to the increased interchain distance by an additional introduction of the CoO_6 octahedra with n in the prism's stead. Structurally, the 1D chain is along the c axis and the triangular lattice is formed in the ab plane in the same structure as $n = 1$. Moreover, the average valence of Co ions that is expressed as $(4n + 2)/(n + 1)$ provided A^{2+} and O^{2-} means that, as n increases from 1, the average Co valence increases from +3 and approaches +4 with increasing n up to ∞. These additional factors about the charge may easily predict the magnetism via complex magnetization processes.

The muon spin rotation and relaxation (μSR) method, as it is very sensitive to the local magnetic environment, is expected to yield crucial information in such a system. Muon particles, discovered by Anderson in 1937, are produced artificially by the decay of pion (π^+), which decays into the muon (μ^+) and the neutrino (ν_μ) with a lifetime of 26 ns:

$$\pi^+ \rightarrow \mu^+ + \nu_\mu. \tag{6.19}$$

The muon is primarily an unstable particle and undergoes a ternary decay into a positron (e^+) and two neutrinos (ν_e, $\bar{\nu}_\mu$) with the lifetime of 2.2 μs like

$$\mu^+ \rightarrow e^+ + \nu_e + \bar{\nu}_\mu. \tag{6.20}$$

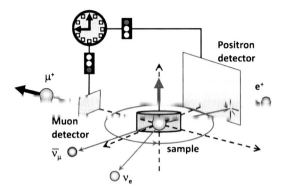

Figure 6.8 Schematic illustration of experimental devices for the wTF-μ^+SR method. *Abbreviation*: wTF, weak transverse field.

In μSR experiments, the polarized muon stays in the sample and we can observe the time variation of the emitted positron. The μSR method, being a microscopic experimental tool, has several characteristics: (i) μ stops at the lattice position and is able to detect the internal field there with high sensitivity; (ii) μSR method makes zero-field measurements possible due to the spin polarization; (iii) μSR technique possesses an original time scale that is a complement to that for neutron spectroscopy and nuclear magnetic resonance (NMR); (iv) this technique is applicable to all the materials in principle and is insensitive to the surface; and (v) μ easily penetrates through the cryostat and the pressure cell.

Here, the weak transverse field (wTF) μ^+SR method was employed, which is useful for detecting the local magnetic order via the shift of the muon spin precession frequency in the applied field and the enhanced muon spin relaxation [10, 11]. Figure 6.8 shows an arrangement view of the experimental equipment. Under the condition that a magnetic field is applied (0, 0, H_0) and the direction of the counter ($\cos\phi$, $\sin\phi$, 0), the wTF-μ^+SR spectrum is given by

$$N_e(t, \phi) = N_e^0 \exp(-t/\tau_\mu)[1 + AP(t)\cos(\phi + \gamma_\mu H_0 t)], \qquad (6.21)$$

where τ_μ, A, and γ_μ are the average lifetime of the muon, the spatial anisotropic parameter (0.2–0.3), and the gyromagnetic ratio ($\gamma_\mu/2\pi = 135.54$ MHz/T), respectively. The exponential part and the term including $P(t)$ represent the relaxation and the rotation of the muon

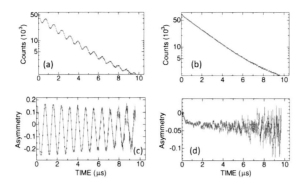

Figure 6.9 wTF-μ^+SR spectra at (a) 193 K and (b) 10 K and its asymmetry at (c) 193 K and (d) 10 K for $Ca_3Co_2O_6$.

spin, respectively. Also, the term in [ ] corresponds to the probability of positron emission. The counters for setup are located at $\phi = 0$ (forward) and π (backward). The spin rotation term is derived by subtracting the count number of each other, namely,

$$\frac{N_e(t, 0) - \alpha N_e(t, \pi)}{N_e(t, 0) + \alpha N_e(t, \pi)} = A P(t)\cos(\gamma H_0 t). \qquad (6.22)$$

Since the above equation shows a difference between signals obtained from symmetric constituted counters, one calls it asymmetry. Here, α is defined as $N_e(0, 0)/N_e(0, \pi)$.

The wTF-μ^+SR spectra and their asymmetry are displayed in Fig. 6.9. The spectrum at 193 K in Fig. 6.9a clearly shows oscillation, whereas that at 10 K does a monotonic decrease with time. The amplitude is also shown in Figs. 6.9c and 6.9d as a time variation, and its temperature dependence is plotted in Fig. 6.10a as the normalized asymmetry (A_{TF}) obtained from the wTF measurements. The first thing that one notices is that the asymmetry, which is proportional to the volume fraction of the paramagnetic phase, decreases below a certain temperature T'_N. It should be noted that there are no marked anomalies in the susceptibility curvemeasured with $H = 10$ kOe in the vicinity of T_N for the four compounds (see Fig. 6.10b). The similar behavior is also observed for other n members and the onset temperatures of T_N are estimated as 100 ± 25 K for $n = 1$, 90 ± 10 K for $n = 2$, 85 ± 10 K for $n = 3$, and 50 ± 10 K for $n = 5$, respectively, and are found to decrease with increasing n (see Fig. 6.11).

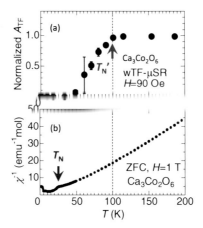

Figure 6.10 Temperature dependences of (a) normalized asymmetry together with (b) inverse susceptibility for $Ca_3Co_2O_6$ [10].

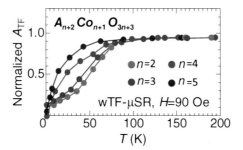

Figure 6.11 Temperature dependence of normalized asymmetry for $n = 2$, 3, 4, and 5 in $A_{n+2}Co_{n+1}O_{3n+3}$ [10].

On the other hand, the zero field (ZF) μ^+SR method to clarify weak local magnetic (dis)order in samples exhibiting quasistatic paramagnetic moments is also performed [10, 11]. As already touched upon, the muon itself is polarized, and thus one can observe the motion change of the muon under a zero external field too. The relationship between the static internal field H_{int} and the frequency ω becomes

$$\omega = \gamma_\mu H_{int}. \tag{6.23}$$

Under the fluctuating field, the magnetic field that the muon detects has a random distribution from the viewpoint of the magnitude and

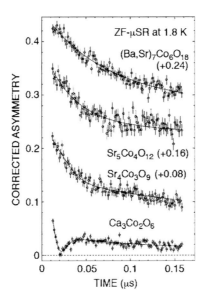

Figure 6.12 ZF-μSR spectra at 1.8 K for $n = 1, 2, 3$, and 5 in $A_{n+2}Co_{n+1}O_{3n+3}$. The spectra are each shifted upward by 0.08 for clarity [10]. *Abbreviation:* ZF, zero field.

the direction. The direction of the internal field is isotropic, and its magnitude and components have the Gaussian distribution in the center of 0:

$$p(H_a) = \frac{1}{\sqrt{2\pi}\sigma_H}\exp\left(-\frac{H_a^2}{2\sigma_H^2}\right) \tag{6.24}$$

and

$$\sigma_H^2 =< H_x^2 >=< H_y^2 >=< H_z^2 >, \tag{6.25}$$

where $a = x$, y, and z. In this case, the relaxation function $G(t)$ defined as $P(t)/P(0)$ is expressed as

$$G(t) = \frac{1}{3} + \frac{2}{3}\left[1 - (\gamma_\mu \sigma_H)^2 t^2\right]\exp\left[-\frac{1}{2}(\gamma_\mu \sigma_H)^2 t^2\right], \tag{6.26}$$

which is called the Kubo–Toyabe function.

With all the decrease in A_{TF} below 100 K for the samples with $n = 2, 3$, and 5, the ZF-μ^+SR spectra exhibit no oscillations even at 1.8 K, as seen in Fig. 6.12. This indicates that the magnetic moment appears but is still fluctuating at 1.8 K. For $n = 1$, on the other hand,

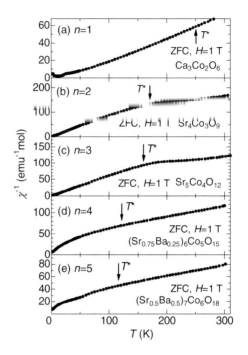

Figure 6.13 Temperature dependence of inverse susceptibility for (a) $n = 1$, (b) $n = 2$, (c) $n = 3$, (d) $n = 4$, and (e) $n = 5$ in $A_{n+2}Co_{n+1}O_{3n+3}$. T^* shows a temperature below which the susceptibility deviates from the Curie–Weiss behavior.

an oscillation is clearly observed, suggesting a long-range magnetic order. For your reference, the ZF-μ^+SR spectra are well fitted by a combination of fast and slow exponential relaxation functions (for fluctuating moments) and a dynamical Kubo–Toyabe function.

Figure 6.13 shows temperature dependence of inverse susceptibility (χ^{-1}) for $n = 1$–5 up to 300 K. χ^{-1} of all the samples deviated downward from the Curie–Weiss law below T^*, implying an FM correlation [12]. The interpretation that this short-range interaction working between the intrachains is natural because the distance between the adjacent Co ions along the CoO_3 chain is smaller than the interchain distance. For example, χ of $n = 1$ changed its slope gradually around 250 K. Hardy et al. reported that magnetic specific heat appears around 200 K and pointed out the presence of short-range ordering [13]. ^{59}Co NMR measurements

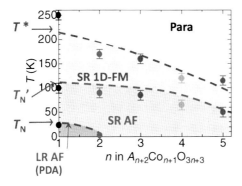

Figure 6.14 Magnetic phase diagram clarified by macroscopic and microscopic techniques. SR 1D-FM, SR AF, and LR AF (PDA) denote short-range 1D ferromagnetism, short-range antiferromagnetism, and long-range partially disordered antiferromagnetism, respectively.

indicate that the intrachain FM correlation steeply develops below 300 K, while the interchain AF one does below 120 K with the critical exponent expected in the 2D Ising antiferromagnet [14]. We left the reason why μSR is unable to observe the short-range 1D FM order untouched but is considered to be because moving solitons along the chain induce a large fluctuation of the internal field at the muon sites. The magnetism clarified by systematic macroscopic and microscopic techniques is summarized in Fig. 6.14 as a phase diagram. From high temperature, a short-range FM ordering develops below T^* due to the intrachain FM interaction. With decreasing temperature, short-range FM spins start to correlate with each other antiferromagnetically and eventually form a PDA state as a result of the long-range AF and FM ordering only for $n = 1$ and 2 (see Fig. 6.15). We may note, in passing, that the PDA transition temperature is enhanced to 90 K for Ca_3CoRhO_6 in which Co in the octahedral site for $Ca_3Co_2O_6$ is replaced with Rh and subsequently a frozen PDA state below 35 K [15].

One's interest may be attracted to the valence and spin state of Co ions, which also draw down the magnetism. Even for $Ca_3Co_2O_6$, the way of combination of the valence in each site is 3: (octahedron, prism) = (Co^{3+}, Co^{3+}), (Co^{2+}, Co^{4+}), and (Co^{4+}, Co^{2+}) despite the average valence is simply $+3$. In general, the

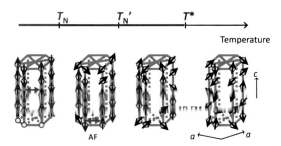

Figure 6.15 Ordering process observed for $Ca_3Co_2O_6$. The arrows with a black and red color are spins without and with specific magnetic interactions, respectively.

spectra of core electrons might yield information on the valence. For XPS measurements, the excitation energy exceeds 1000 eV; thus core electrons having large binding energies can also be excited. The correspondence of the actual electronic structure and the photoemission spectra is illustrated in Fig. 6.16. When a monochromatized undulator beam with $h\nu$ is radiated on a sample in vacuum, the binding energy of the electrons in materials can be determined by measuring the kinetic energy E_k of the photoelectron

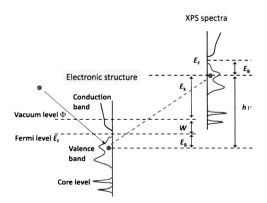

Figure 6.16 Relationship between the electronic structure and the XPS spectra. $h\nu$, E_k, W, and E_b are the incidence energy, the kinetic energy, the work function, and the binding energy, respectively. *Abbreviation*: XPS, X-ray photoemission spectroscopy.

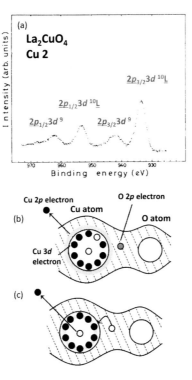

Figure 6.17 (a) Cu $2p$ core-level XPS spectra for La_2CuO_4[16] and screening processes after photoemission, (b) $\underline{2p}3d^9$, and (c) $\underline{2p}3d^{10}\underline{L}$.

and becomes

$$E_b = h\nu - E_k - W, \tag{6.27}$$

using the work function W. A monochromatized radiation impinges on the surface of the sample kept in a high vacuum of $\leq 10^{-9}$ torr. In this way the XPS is the surface-sensitive probe; the contamination on the surface should be removed by filing, cleaving, or sputtering. The existence of adsorbed CO and CO_2 is detectable by investigating the spectra at 284 eV (C $1s$) and 531 eV (O $1s$). Please note here that O is, of course, included in oxides, but the peak position is different from that of CO or CO_2.

Next, how the core-level spectra vary depending on the valence is explained taking La_2CuO_4 as an example. Figure 6.17 shows the Cu $2p$ core-level spectrum [16]. The $2p$ orbital splits into two

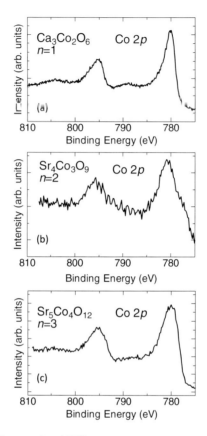

Figure 6.18 Co $2p$ core-level XPS spectra at room temperature for (a) $n = 1$, (b) $n = 2$, and (c) $n = 3$.

levels ($j = 1/2, 3/2$) due to the spin–orbit interaction, for example, $2p_{1/2}3d^{10}\underline{L}$ and $2p_{3/2}3d^{10}\underline{L}$. However, the number of the component of the Cu $2p$ core-level spectrum does not match two but four. As shown in Figs. 6.17b and 6.17c, when the hole is produced in the Cu $2p$ level due to the photoelectric effect, the screening effect gives rise to the rearrangement of valence electrons. As a final state, there are two cases where the hole exists in Cu atom and O atom: Fig. 6.17b is corresponding to the situation where the Cu $2p$ level is screened by the O $2p$ electrons spatially spread over widely, while in the case of Fig. 6.17c the Cu $2p$ level is screened by itself by being supplied electrons from the O $2p$ level. These states are denoted as $2\underline{p}3d^9$ and

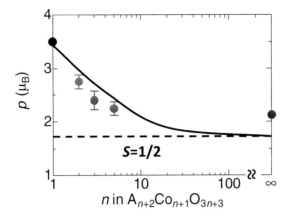

Figure 6.19 Effective magnetic moment vs n in $A_{n+2}Co_{n+1}O_{3n+3}$. The dashed line represents the effective magnetic moment (1.73 μ_B) corresponding to $S = 1/2$ (LS, Co^{4+}). The solid line is a guide for eyes [10].

$2p3d^{10}\underline{L}$, respectively. The presence of two components even for the same j ($= 1/2$) demonstrates the initial state of Cu $3d^9$ in La$_2$CuO$_4$.

As for the relationship between the Co $2p$ core-level spectrum and the valence, the data for Co^{2+}, Co^{3+}, and Co^{4+} is already well established [17, 18]. The difference among them appears in the position of the satellite peak, that is, 787 eV with relatively high intensity for Co^{2+}, 789 eV with low intensity for Co^{3+}, and inexistent for Co^{4+}. Figure 6.18a shows the Co $2p$ core-level spectrum for Ca$_3$Co$_2$O$_6$. The weak satellite structure is found at 789 eV, which indicates that both Co ions in the octahedral and prismatic sites are Co^{3+}. With increasing n, the satellite structure is found to be less visible (see Fig. 6.18b,c). This indicates that Co ions are in the mixed states of Co^{3+} and Co^{4+}.

To confirm the nature of the 2D AF interaction, the physical pressure effect on T'_N was investigated by conducting μSR measurements under pressures of up to 1.1 GPa. A larger pressure dependence on T'_N is noticeably observed as n is increased, for example, dT'_N/dP = 2.2 K/GPa for $n = \infty$. The interchain (d_{ic}) variation of T'_N is well fitted by $T'_N/T'_{N,0} = (1 - d_{ic}/d_{ic,0})^\beta$, where $\beta = 0.307$ and $d_{ic,0} = 5.65$ Å. This analysis means that the slope of the $T'_N(d)$ curve is steeper in the vicinity of 5.65 Å, and therefore

Table 6.1 Composition, material, and nominal charge/spin distribution for $A_{n+2}Co_{n+1}O_{3n+3}$

n	Materials	Prismatic site	Octahedral site
1	$Ca_3Co_2O_6$	Co^{3+} (HS)	Co^{3+} (LS)
2	$Sr_4Co_3O_9$	Co^{3+} (HS)	Co^{3+} (LS), Co^{4+} (LS)
3	$Sr_5Co_4O_{12}$	Co^{3+} (HS)	Co^{3+} (LS), Co^{4+} (LS), Co^{4+} (LS)
5	$(Sr_{0.5}Ba_{0.5})_7Co_6O_{18}$	Co^{3+} (HS)	Co^{3+} (LS), Co^{4+} (LS), Co^{4+} (LS), Co^{4+} (LS)
∞	$BaCoO_3$	—	Co^{4+} (LS)

the shrinkage of the lattice gives much decrease in T'_N for larger n, which agrees with our pressure results. The universal $T'_N(d)$ and the dT'_N/dP behavior clearly suggest the existence of the short-range 2D AF order, which is partly driven by the short-range 1D FM order below T^* and forward to the long-range AF order below T_N.

The effective magnetic moment of Co ions is obtained by fitting the $\chi(T)$ curve between 300 and 600 K using the Curie–Weiss law. As a concrete calculation procedure, the effective magnetic moment is sought as

$$p = \sqrt{\frac{3k_B C_{mol}}{N_{mol}\mu_B^2}}, \tag{6.28}$$

where C_{mol} and N_{mol} are the Curie constant and the Avogadro number per 1 mol. The possible valence and the spin state are summarized in Table 6.1. The Co ion in the prismatic site is trivalent with a high-spin (HS) state ($S = 2$), whereas Co ions in the octahedral sites are trivalent and tetravalent with a low-spin (LS) state. As n is increased, it can be interpreted that an LS Co^{4+} ($S = 1/2$) is newly added. Figure 6.19 displays the n variation of the effective magnetic moment (p) derived from the magnetization data. The $p(n)$ curve is found to obey the relationship of

$$p^2 = \frac{1}{n+1}p_{S=2}^2 + \frac{n-1}{n+1}p_{S=1/2}^2 + \frac{1}{n+1}p_{S=0}^2. \tag{6.29}$$

Namely, Co ions in the 1D chain consisting of $(n + 1)$ sites are made of one HS Co^{3+} ($S = 2$) in the trigonal prism, $(n − 1)$ LS Co^{4+} ($S = 1/2$) in the octahedron, and one LS Co^{3+} ($S = 0$) in the octahedron. In retrospect, the systematic decrease in p perhaps causes the decrease in T^*.

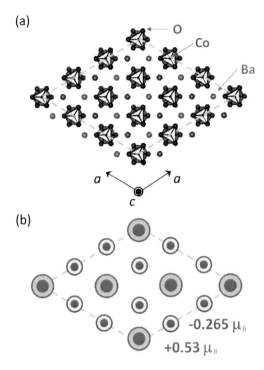

(a)

(b)

-0.265 μ_B

+0.53 μ_B

Figure 6.20 (a) Crystal structure (c plane view) and (b) magnetic structure formed below 15 K for BaCoO$_3$. Co moments align parallel (+) or antiparallel (−) to the c axis with an amplitude modulation [19].

The precise magnetic structure analysis for hexagonal perovskite BaCoO$_3$ (A = Ba, $n = \infty$) by neutron diffraction and μSR in the temperature range between 60 K and 1.5 K revealed the complex magnetism [19]. Pure Co^{4+} is known to be unstable as against Co^{2+} and Co^{3+}, and thereby the single-phased sample was obtained by annealing at 650°C for 150 hours in oxygen under 1 MPa pressure. The magnetic phase diagram of this material does not link that for $n \leq 5$; that is, the 1D FM order along the CoO$_3$ chain is completed below 53 K, and then each chain acts as a single spin so as to form the 2D AF order categorized into a modulated AF Γ_3 structure rather than the well-known classical AF 120° structure (see Fig. 6.20). The magnitude of the Co moments antiparallel to the c axis is half of that parallel to the c axis.

References

1. Brisi C, Rolando P (1968), Ricerche sul sistema ossido di calcio- ossido cobaltoso-ossigeno, *Ann. Chim. (Rome)*, **58**, 676–683.
2. Aasland S, Fjellvåg H, Hauback B (1997), Magnetic properties of the one-dimensional $Ca_3Co_2O_6$, *Solid State Commun.*, **101**, 187–192.
3. Maignan A, Michel C, Masset AC, Martin C, Raveau B (2000), Single crystal study of the one dimensional $Ca_3Co_2O_6$ compound: fieve stable configurations for the Ising triangular lattice, *Euro. Phys. J. B*, **15**, 657–663.
4. Mohapatra N, Iyer KK, Das SD, Chalke BA, Purandare SC, Sampathku-maran EV (2009), Magnetic behavior of nanocrystals of the spin-chain system $Ca_3Co_2O_6$: absence of multiple steps in the low-temperature isothermal magnetization, *Phys. Rev. B*, **79**, R140409-1–R140409-4.
5. Takami T, Horibe M, Itoh M, Cheng J (2010), Controlling independently the electric and thermal properties by shrinking the particle size down to nanosize in quasi-one-dimensional $Ca_3Co_2O_6$ and $Sr_6Co_5O_{15}$, *Phys. Rev. B*, **82**, 085110-1–085110-6.
6. Mekata M (1977), Antiferro-ferrimagnatic transition in triangular Ising lattice, *J. Phys. Soc. Jpn.*, **42**, 76–82.
7. Achiwa N (1969), Linear antiferromagnetic chains in hexagonal $ABCl_3$-type compounds (A; Cs, or Rb, B; Cu, Ni, Co, or Fe), *J. Phys. Soc. Jpn.*, **27**, 561–574.
8. Boulahya K, Parras M, Gonza Llez-Calbet JM (1999), The $A_{n+2}B_nB'O_{3n+3}$ family ($B = B' = Co$): ordered intergrowth between $2H$-$BaCoO_3$ and $Ca_3Co_2O_6$ structures, *J. Solid State Chem.*, **145**, 116–127.
9. Yamaura K, Zandbergen HW, Abe K, Cava RJ (1999), Synthesis and properties of the structurally one-dimensional cobalt oxide $Ba_{1-x}Sr_xCoO_3$ ($044 \le x \le 0.5$), *J. Solid State Chem.*, **146**, 96–102.
10. Sugiyama J, Nozaki H, Brewer JH, Ansaldo EJ, Takami T, Ikuta H, Mizutani U (2005), Appearance of a two-dimensional antiferromagnetic order in quasi-one-dimensional cobalt oxides, *Phys. Rev. B*, **72**, 064418-1–064418-11.
11. Sugiyama J, Nozaki H, Ikedo Y, Mukai K, Andreica D, Amato A, Brewer JH, Ansaldo EJ, Morris GD, Takami T, Ikuta H (2006), Evidence of two dimensionality in quasi-one-dimensional cobalt oxides, *Phys. Rev. Lett.*, **96**, 197206-1–197206-4.

12. Takami T, Nozaki H, Sugiyama J, Ikuta H (2007), Magnetic properties of one-dimensional compounds $A_{n+2}Co_{n+1}O_{3n+3}$ ($A = $ Ca, Sr, Ba; $n = $ 1–5), *J Magn. Magn. Mater.*, **310**, e438–e440.

13. Hardy V, Lambert S, Lees MR, Paul DM (2003), Specific heat and magnetization study on single crystals of the frustrated quasi-one-dimensional oxide $Ca_3Co_2O_6$, *Phys. Rev. B*, **68**, 014424-1–014424-7.

14. Shimizu Y, Horibe M, Nanba H, Takami T, Itoh M (2010), Anisotropic spin dynamics in the frustrated chain $Ca_3Co_2O_6$ detected by single-crystal ^{59}Co NMR, *Phys. Rev. B*, **82**, 094430-1–094430-8.

15. Niitaka S, Yoshimura K, Kosuge K, Nishi M, Kakurai K (2001), Partially disordered antiferromagnetic phase in Ca_3CoRhO_6, *Phys. Rev. Lett.*, **87**, 177202-1–177202-4.

16. Takahashi T, Maeda F, Katayama-Yoshida H, Okabe Y, Suzuki T, Fujimori A, Hosoya S, Shamoto S, Sato M (1988), Photoemission study of single-crystalline $(La_{1-x}Sr_x)_2CuO_4$, *Phys. Rev. B*, **37**, 9788–9791.

17. van Elp J, Wieland JL, Eskes H, Kuiper P, Sawatzky GA, de Groot FMF, Turner TS (1991), Electronic structure of CoO, Li-doped CoO, and $LiCoO_2$, *Phys. Rev. B*, **44**, 6090–6103.

18. Saitoh T, Mizokawa T, Fujimori A, Abbate M, Takeda Y, Takano M (1997), Electronic structure and magnetic states in $La_{1-x}Sr_xCoO_3$ studied by photoemission and x-ray-absorption spectroscopy, *Phys. Rev. B*, **56**, 1290–1295.

19. Nozaki H, Janoschek M, Roessli B, Sugiyama J, Keller L, Brewer JH, Ansaldo EJ, Morris GD, Takami T, Ikuta H (2007), Neutron diffraction and μSR study on the antiferromagnet $BaCoO_3$, *Phys. Rev. B*, **76**, 014402-1–014402-7.

Chapter 7

Superconductivity

7.1 Prologue

Lattice vibrations in metals generally give rise to a finite electrical resistivity, and it disappears only when the metal resumes perfectly periodic ion potentials at absolute zero temperature. But there are a number of metals the electrical resistivity of which vanishes at finite temperatures together with the Meissner effect, Josephson effect, and flux quantum. These basic properties are distinctive for superconductivity. Fifty-seven elements are currently known to exhibit superconductivity. To take a simple example, superconductivity under high pressure was observable at extremely low temperature, even for the familiar gas oxygen (see Fig. 7.1) [1]. Historically, superconductivity for mercury after liquefying helium was discovered in 1911. Since then, superconductivity has been actively studied in many metals, alloys, and oxides. The mechanism of superconductivity had remained unresolved for many years as one of the most mysterious puzzles in physics until Bardeen, Cooper, and Schrieffer (BCS) put forward the epoch-making theory, that is, BCS theory, in 1957 [2]. They established the fundamentals on the basis of the Cooper pair–mediated electron–phonon interaction. According to their theory, the maximum superconductive transition

Functional Cobalt Oxides: Fundamentals, Properties, and Applications
Tsuyoshi Takami
Copyright © 2014 Pan Stanford Publishing Pte. Ltd.
ISBN 978-981-4463-32-4 (Hardcover), 978-981-4463-33-1 (eBook)
www.panstanford.com

temperature (T_c) has been considered to be 30–40 K at most. In 1986, however, Bednorz and Müller reported the possibility of superconductivity for La-Ba-Cu-O and won the Novel Prize in physics in 1987 [3]. Motivated by their work, researchers in solid-state physics have taken some important steps in high-T_c super-conductors for cuprates. As a theoretical concept proposed soon after the discovery, resonating valence band (RVB) theory suggested by Anderson is famous, in which doublons and holons form bound charge-neutral excitations and lead to zero electrical resistivity, including, but not limited to, the square and triangular lattices [4]. In Section 7.1, Bose–Einstein condensation is introduced, which would help one's interpretation of superconductivity. In Section 7.2, high-T_c cuprate superconductors and digest applications are introduced historically. Superconductivity is also reported for other transition metal oxides; superconductivity observed for water-intercalated $Na_x CoO_2$ is discussed as a first example for cobalt oxide superconductors in Section 7.3.

7.2 Bose–Einstein Condensation

Let us start with two identical particles that have the same mass, charge, and spin. When their positions r_1 and r_2 are exchanged, the relationship between the wave function φ is written as

$$\varphi(r_1, r_2) = c\varphi(r_2, r_1), \qquad (7.1)$$

where c is a constant operator. One more similar procedure brings about

$$\varphi(r_1, r_2) = c^2\varphi(r_1, r_2). \qquad (7.2)$$

Thus, the condition $c = 1$ or -1 is obtained. The particles with $c = 1$ and -1 are called the Bose particle and the Fermi particle, respectively. In general, the photon belongs to the former group and the electron, proton, and neutron to the latter one. However, the electron exhibits both natures under special conditions. Once Bose–Einstein condensation takes place, the wave function Ψ of N Fermi particles, bearing in mind the spin quantum number σ, is expressed

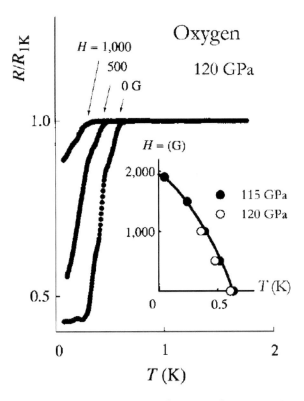

Figure 7.1 Magnetic field dependence of superconductive transition at 120 GPa. Inset shows a critical magnetic field versus temperature diagram of oxygen at 115 GPa (filled circles) and 120 GPa (open circles) [1].

as

$$\Psi(\mathbf{r}_1, \mathbf{r}_2, \cdots, \mathbf{r}_N; \sigma_1, \sigma_2, \cdots, \sigma_N) =$$
$$\mathcal{A}[\varphi(\mathbf{r}_1, \mathbf{r}_2; \sigma_1, \sigma_2)\varphi(\mathbf{r}_3, \mathbf{r}_4; \sigma_3, \sigma_4)\cdots\varphi(\mathbf{r}_{N-1}, \mathbf{r}_N; \sigma_{N-1}, \sigma_N)], \quad (7.3)$$

where $\mathcal{A}$ represents the switch of coordination and spin of N electrons as well as normalization. It must be noted that all the electron pairs are described by the same φ. Ψ in Eq. 7.3 is referred to as the BCS wave function. Details in BCS theory are too involved to be treated here. Bose–Einstein condensation is also reported for superfluidity of ^{4}He and ^{3}He, the alkaline atom, and the neutron star.

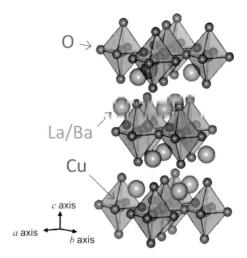

O →

La/Ba

Cu

c axis

a axis

b axis

Figure 7.2 Crystal structure of $La_{2-x}Ba_xCuO_4$.

7.3 High-T_c Cuprate Superconductors

Bednorz and Müller discovered superconductivity in the La-Ba-Cu-O system in 1986 [3], which was found to possess the composition of $La_{2-x}Ba_xCuO_4$ later [6]. Its maximum T_c was 30 K, but the enhancement of T_c was actively reported. The partial substitution of La^{3+} for Ba^{2+} means the introduction of holes to Cu^{2+}, the average valence of which is $2 + x$. The system has the K_2NiF_4 structure with a single CuO_2 sheet separated by rock-salt layers consisting of two La/BaO sheets (see Fig. 7.2). The rock-salt layer is not a charge reservoir, which makes the fractions of holes per Cu atom in the CuO_2 layer unambiguously given by the Ba concentration x, provided oxygen stoichiometry is maintained.

Since the discovery of high-T_c superconductivity, the value of which surpassed the past maximum T_c of 23 K for the Nb_3Ge compound [5], this one of the most exciting and challenging topics has attracted much attention from fundamental and practical points of view. Wu et al. have reported $T_c = 92$ K above the liquid nitrogen temperature for $YBa_2Cu_3O_y$ with a complex crystal structure [7]. Thereafter, T_c shows a rising trend: 110 K for Bi-Sr-Ca-Cu-O [8], 125 K for Tl-Ba-Ca-Cu-O [9], and 135 K (164 K under high pressure) for

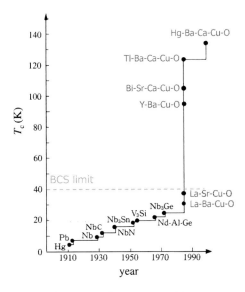

Figure 7.3 Progress of superconducting transition temperature with the times.

Hg-Ba-Ca-Cu-O [10, 11]. The historical variation of T_c is summarized in Fig. 7.3.

In addition to the fundamental research on the mechanism of high-T_c superconductivity, high-T_c superconductors have been studied from an application side since liquid nitrogen can be used as a freezing mixture. For application as electric power among various fields, the high values of the critical current density (J_c) and the critical magnetic field (H_c) are sought as well as a high-T_c. The application area spreads over many divergences such as energy, medicine, electronics, and transport. Superconducting magnet energy storage (SMES) can reserve power energy as magnetic energy by applying a current in the coil made of a superconductor. Power energy can be also available effectively without waste by transmission using a superconducting cable. A magnetic levitation train, magship, magnetic resonance imaging, being a benefit of the technology in the field of superconductors, is also well known to the general public. The pinning effect is applicable to superconducting permanent magnet materials.

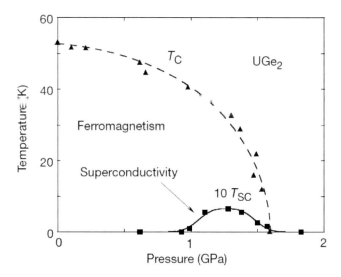

Figure 7.4 Temperature–pressure phase diagram of UGe$_2$. T_c denotes the Curie temperature and T_{sc} the superconducting transition temperature [15].

Recently, research on superconductivity is wildly extended again by the discovery of superconductivity for the Fe-As system [12]. Especially one should pay attention to its composition element, to sum up, the magnetic Fe element. Compounds containing Fe must not exhibit Superconductivity, since magnetism and conductivity are opposite aspects of each other, and thereby the conventional sense is discredited by the discovery of superconductivity for Fe compounds such as RFeAs(O,F) and AFe$_2$As$_2$ (R: the rare-earth element, A: the alkaline-earth metal) [13]. In the structural sense, AFe$_2$As$_2$ is a layered system similar to the case of cuprate superconductors. As a diversity, the mother material is the Mott insulator, whereas that for Fe-based compounds shows a metallic behavior. Furthermore, the Fermi surface is composed of Cu$3d$ and O$2p$ orbitals, but the contribution mainly from Fe$3d$ orbitals rather than from anions should be taken into account. In contrast to the general trend that impurity doping is known to obstruct superconductivity, for instance, Co doping to Fe-based compounds does not reduce their T_c so significantly [14].

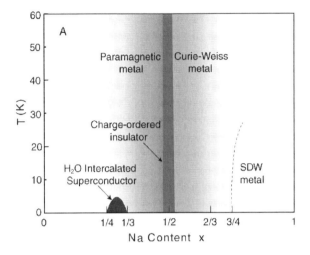

Figure 7.5 Phase diagram of $Na_x CoO_2$ [18].

Being associated with ferromagnetism, superconductivity was uncovered below the ferromagnetic temperature for UGe_2, and its phase diagram is shown in Fig. 7.4 [15]. Although cuprate superconductors are typical for strongly correlated systems, the first example of a superconductor in this category is $CeCu_2Si$ [16].

7.4 Superconductivity of $Na_x CoO_2 \cdot yH_2 O$

In this section, I will shift the emphasis away from cuprates to cobalt oxides. Since the discovery of a huge thermoelectric power for $x = 0.5$ in $Na_x CoO_2$ [17], research on the system with varying x has been accelerated. The phase diagram looks something like Fig. 7.5. As x is increased from 0.3, the ground state goes from a paramagnetic metal to a Curie–Weiss metal and lastly to a spin-density wave metal *via* a charge-ordered insulator at $x = 0.5$ [18]. Let's take a Curie–Weiss metal example, which means that the localized moment does not necessarily suggest typical Curie–Weiss behavior. One scenario is the coexistence of the magnetic phase with localized spins embedded in a matrix with an itinerant character, as proposed by several reports [19].

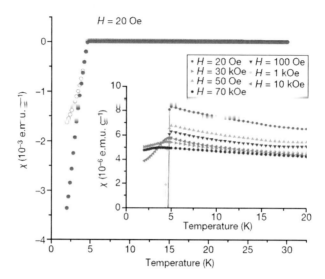

Figure 7.6 Temperature dependence of magnetic susceptibility for $Na_xCoO_2 \cdot yH_2O$ measured under the zero-field cooling (filled) and field-cooling (open) modes [20].

The superconducting state with $T_c = 5$ K appears by intercalation with H_2O for Na_xCoO_2 (see Fig. 7.6) [20], and it spreads over the compositions of $x = 1/4$–$1/3$ in $Na_xCoO_2 \cdot yH_2O$ [21]. Intercalation is generally the insertion of a molecule between two other molecules. T_c of this compound displays the same kind of behavior with curvature on chemical doping that is observed in the high-T_c copper oxides. The control of Na content mainly means changes of the c axis length; The c/a ratio actually increased to be 6.95 for $Na_{0.35}CoO_2 \cdot 1.3H_2O$, being much larger than 3.97 of that without water. There is a marked resemblance in superconducting properties between $Na_xCoO_2 \cdot yH_2O$ and high-T_c copper oxides, suggesting that the two systems have similar underlying physics. In the high-T_c copper oxides, superconductivity occurs in the CuO_2 square lattice when the hole is doped in the Mott insulator, in which the Cu^{2+} ($S = 1/2$) moments are antiferromagnetically ordered in the CuO_2 plane, whereas the present superconductivity can be understood by electron doping of approximately 0.35 per Co atom to a low-spin Co^{4+} ($S = 1/2$) in the CoO_2 triangular lattice. The optimal doping level for superconductivity is clearly higher with

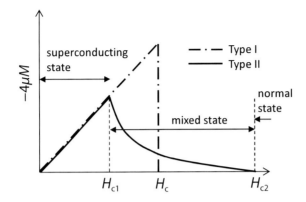

Figure 7.7 Equilibrium magnetization curve of a type II superconductor together with that of a type I superconductor for comparison.

respect to the Mott–Hubbard-like half-filled two-electron band in $Na_xCoO_2 \cdot 1.3H_2O$ (0.3 electrons) than in the copper oxides (0.15 electrons or holes).

This system is a type II superconductor, in which the mixed state between the normal state and the Meissner state exists for $H_{c1} \leq H \leq H_{c2}$ below T_c, where H_{c1} and H_{c2} are the lower critical field and upper critical field, respectively, as opposed to a type I superconductor characterized by a single critical H_c at the boundary of the superconducting and normal conducting states (see Fig. 7.7). The magnetic susceptibility shows a strong temperature variation even above T_c [22]. The spin–spin relaxation rate $1/T_1$ measured by nuclear magnetic resonance (NMR) experiments decreases proportionally T^3 without showing a coherent peak just below T_c, indicative of the presence of the node [23]. Furthermore, the Knight shift does not depend on a temperature below T_c for the muon spin rotation and relaxation (μSR) measurements, which provides a possibility of a spin triplet superconductor [24]. In contrast to the spin triplet picture, the data of the photoemission Spectroscopy [25] and inelastic neutron scattering and Knight shift in NMR measurements suggest a spin singlet [26]. On the other hand, the existence of an f wave spin triplet is proposed [27]. As other insights, an s wave spin singlet model or the conventional BCS model mediated by the electron–phonon interaction is predicted

[28]. Being similar to the band structure of $Na_x CoO_2$, the d_{xy} band has a local minimum structure at the Γ point, and thus a double Fermi surface is formed with the center at the Γ point. The nesting of these Fermi surfaces implies the existence of a reciprocal lattice vector that is called the scattering vector, which corresponds to a long-range magnetic order in real space. Kuroki et al. pointed out the development of spin fluctuation due to the nesting between two Fermi surfaces and then the occurrence of the unconventional extended *s* wave superconductivity through the medium of the fluctuation [29]. In this way, there is room for argument on this point.

References

1. Shimizu K, Suhara K, Ikumo M, Eremets MI, Amaya K (1998), Superconductivity in oxygen, *Nature*, **393**, 767–769.

2. Bardeen J, Cooper LN, Schrieffer JR (1957), Theory of superconductivity, *Phys. Rev. Lett.*, **108**, 1175–1204.

3. Bednorz JG, Müller KA (1986), Ricerche sul sistema ossido di calcio-ossido cobaltoso-ossigeno, *Z. Phys.*, **B64**, 189–193.

4. Anderson PW (1987), The resonating valence bond state in $La_2 CuO_4$ and superconductivity, *Science*, **235**, 1196–1198.

5. Matthias BT, Geballe TH, Longinotti LD, Corenzwit E, Hull GW, Willens RH, Maita JP (1967), Superconductivity at 20 degrees Kelvin, *Science*, **156**, 645–646.

6. Uchida S, Takagi H, Tanaka S, Kitazawa K (1987), High T_c superconductivity of La-Ba-Co oxides, *Jpn. J. Appl. Phys.*, **26**, L1–L2.

7. Wu MK, Ashburn JR, Torng CJ, Hor PH, Meng RL, Gao L, Huang ZJ, Wang YQ, Chu CW (1987), Superconductivity at 93 K in a new mixed-phase Y-Ba-Cu-O compound system at ambient pressure, *Phys. Rev. Lett.*, **58**, 908–910.

8. Maeda H, Tanaka Y, Fukutomi M, Asano T (1988), A new high-T_c oxide superconductor without a rare earth element, *Jpn. J. Appl. Phys.*, **27**, L209–L210.

9. Sheng ZZ, Hermann AM (1988), Bulk superconductivity at 120 K in the Tl-Ca/Ba-Cu-O system, *Nature*, **332**, 138–139.

10. Schilling A, Cantoni M, Guo JD, Ott HR (1993), Superconductivity above 130 K in the Hg-Ba-Ca-Cu-O system, *Nature*, **363**, 56–58.

11. Gao L, Xue YY, Chen F, Xiong Q, Meng RL, Ramirez D, Chu CW (1994), Superconductivity up to 164 K in $HgBa_2Ca_{m-1}Cu_mO_{2m+2+\delta}$ ($m = 1, 2$, and 3) under quasihydrostatic pressures, *Phys. Rev. B*, **50**, 4260–4263.

12. Kamihara Y, Watanabe T, Hirano M, Hosono H (2008), Iron-based layered superconductor $La[O_{1-x}F_x]FeAs$ ($x = 0.05$-0.12) with $T_c = 26$ K, *J. Am. Chem. Soc.*, **130**, 3296–3297.

13. Shelton RN, Braun HF, Musick E (1984), Superconductivity and relative phase stability in 1:2:2 ternary transition metal silicides and germanides, *Solid State Commun.*, **52**, 797–799.

14. Wang C, Li YK, Zhu ZW, Jiang S, Lin X, Luo YK, Chi S, Li LJ, Ren Z, He M, Chen H, Wang YT, Tao Q, Cao GH, Xu ZA (2009), Effects of cobalt doping and phase diagrams of $LFe_{1-x}Co_xAsO$ ($L =$ La and Sm), *Phys. Rev. B*, **79**, 054521-1–054521-9.

15. Saxena SS, Agarwal P, Ahilan K, Grosche FM, Haselwimmer RKW, Steiner MJ, Pugh E, Walker IR, Julian SR, Monthoux P, Lonzarich GG, Huxley A, Sheikin I, Braithwaite D, Flouquet J (2000), Superconductivity on the border of itinerant-electron ferromagnetism in UGe_2, *Nature*, **406**, 587–592.

16. Steglich F, Aarts J, Bredl CD, Lieke W, Meschede D, Franz W, Schäfer H (1979), Superconductivity in the presence of strong Pauli paramagnetism: $CeCu_2Si_2$, *Phys. Rev. Lett.*, **43**, 1892–1896.

17. Terasaki I, Sasago Y, Uchinokura K (1997), Large thermoelectric power in $NaCo_2O_4$ single crystals, *Phys. Rev. B*, **56**, R12685–R12687.

18. Foo ML, Wang Y, Watauchi S, Zandbergen HW, He T, Cava RJ, Ong NP (2004), Charge ordering, commensurability, and metallicity in the phase diagram of the layered Na_xCoO_2, *Phys. Rev. Lett.*, **92**, 247001-1–247001-4.

19. For example, Carretta P, Mariani M, Azzoni CB, Mozzati MC, Bradarić I, Savić I, Feher A, Šebek J (2004), Mesoscopic phase separation in Na_xCoO_2 ($0.65 \leqslant 0 \leqslant 0.75$), *Phys. Rev. B*, **70**, 024409-1–024409-9.

20. Takada K, Sakurai H, Takayama-Muromachi E, Izumi F, Dilanian RA, Sasaki T (2003), Superconductivity in two-dimensional CoO_2 layers, *Nature*, **422**, 53–55.

21. Schaak RE, Klimczuk T, Foo ML, Cava RJ (2003), Superconductivity phase diagram of $Na_xCoO_2 \cdot 1.3H_2O$, *Nature*, **423**, 527–529.

22. Sakurai H, Takada K, Yoshii S, Sasaki T, Kindo K, Takayama-Muromachi E (2003), Unconventional upper- and lower-critical fields and normal-

state magnetic susceptibility of the superconducting compound $Na_{0.35}CoO_2 \cdot 1.3H_2O$, *Phys. Rev. B*, **68**, 132507-1–132507-3.

23. Ishida K, Ihara Y, Maeno Y, Michioka C, Kato M, Yoshimura K, Takada K, Sasaki T, Sakurai H, Takayama-Muromachi E (2003), Unconventional superconductivity and nearly ferromagnetic spin fluctuations in Na₀.₃₅CoO₂·₁.₃H₂O, *J. Phys. Soc. Jpn.*, **72**, 3041–3044.

24. Higemoto W, Ohishi K, Koda A, Saha SR, Kadono R, Ishida K, Takada K, Sakurai H, Takayama-Muromachi E, Sasaki T (2004), Possible unconventional superconductivity in $Na_xCoO_2 \cdot yH_2O$ probed by muon spin rotation and relaxation, *Phys. Rev. B*, **70**, 134508-1–134508-5.

25. Shimojima T, Ishizaka K, Tsuda S, Kiss T, Yokoya T, Chainani A, Shin S, Badica P, Yamada K, Togano K (2006), Angle-resolved photoemission study of the cobalt oxide superconductor $Na_xCoO_2 \cdot yH_2O$: observation of the Fermi surface, *Phys. Rev. Lett.*, **97**, 267003-1–267003-4.

26. Kobayashi Y, Yokoi M, Sato M (2003), [59]Co-NMR Knight Shift of superconducting $Na_xCoO_2 \cdot yH_2O$, *J. Phys. Soc. Jpn.*, **72**, 2453–2455.

27. Mochizuki M, Ogata M (2006), Deformation of electronic structures due to CoO_6 distortion and phase diagrams of $Na_xCoO_2 \cdot yH_2O$, *J. Phys. Soc. Jpn.*, **75**, 113703-1–113703-5.

28. Yada K, Kontani H (2006), Electron-phonon mechanism for superconductivity in $Na_{0.35}CoO_2$: valence-band Suhl-Kondo effect driven by Shear phonons, *J. Phys. Soc. Jpn.*, **75**, 033705-1–033705-4.

29. Kuroki K, Onari S, Tanaka Y, Arita R, Nojima T (2006), Extended s-wave pairing originating from the a_{1g} band in $Na_xCoO_2 \cdot yH_2O$: single-band U-V model with fluctuation exchange method, *Phys. Rev. B*, **73**, 184503-1–184503-6.

Chapter 8

Transport Properties Combined with Charge, Spin, and Orbital: Magnetoresistance and Spin Blockade

8.1 Prologue

Magnetoresistance was a physical phenomenon found by Thomson in 1856 [1]. The electrical resistivity is varied by the external magnetic field. Its percentage was only a few percent in those days, but Gruenberg and Fert discovered giant magnetoresistance (GMR) for the Cr-Fe film and won the Nobel Prize in physics in 2007 [2, 3]. Motivated by this achievement, it has been clarified that a multilayer film shows the GMR effect, in which ferromagnetic (FM) metal and nonmagnetic layers are stacked by turns. The magnetic head using a more drastic magnetoresistive effect, that is, a colossal magnetoresistive (CMR) effect observed in manganese oxides extended to the recent development of the high-density hard disk. In the first half of this chapter, magnetoresistive effects, including conventional effects, GMR, and CMR, are explained— $R\text{BaCo}_2\text{O}_{5+\delta}$ (R: the rare-earth element) is introduced as one of the examples that show a relatively large magnetoresistance among Co

Functional Cobalt Oxides: Fundamentals, Properties, and Applications
Tsuyoshi Takami
Copyright © 2014 Pan Stanford Publishing Pte. Ltd.
ISBN 978-981-4463-32-4 (Hardcover), 978-981-4463-33-1 (eBook)
www.panstanford.com

oxides. In addition, the extremely large magnetoresistance reported for PdCoO$_2$ is presented.

In the second half, the spin blockade phenomenon that prevents charge from moving freely due to the spatial distribution of spins is introduced. This phenomenon itself was uncovered and experimented on in the late 1980s as a microscopic property. This time, the fact that the spin blockade occurs originating from strong electron correlation in the bulk and dominates macroscopic properties must constantly be borne in mind.

8.2 The Magnetoresistance Effect

When we have two types of carriers, the current density under the magnetic field is given by

$$\mathbf{J} = \left(\frac{\sigma_1}{1 + \beta_1^2 H^2} + \frac{\sigma_2}{1 + \beta_2^2 H^2} \right) \mathbf{E} - \left(\frac{\sigma_1 \beta_1}{1 + \beta_1^2 H^2} + \frac{\sigma_2 \beta_2}{1 + \beta_2^2 H^2} \right) \mathbf{H} \times \mathbf{E}$$

$$(8.1)$$

in the condition such that $\mathbf{J} \perp \mathbf{B}$ and $\mathbf{E} \perp \mathbf{B}$, where σ_i and β_i, $\mathbf{H}$, and $\mathbf{E}$ are, respectively, the conductivity of the ith carrier, $q_i \tau_i / m_i$, the magnetic field, and the electric field. The ratio of the current to electric field is written as

$$\sigma = J^2 / \mathbf{J} \cdot \mathbf{E}. \qquad (8.2)$$

Thus, the conductivity under the magnetic field for the two-band model is

$$\sigma = \frac{\sigma_1/(1 + \beta_1^2 H^2) + \sigma_2/(1 + \beta_2^2 H^2) + \beta_1 H \sigma_1/(1 + \beta_1^2 H^2) + \beta_2 H \sigma_2/(1 + \beta_2^2 H^2)}{\sigma_1/(1 + \beta_1^2 H^2) + \sigma_2/(1 + \beta_2^2 H^2)}.$$

$$(8.3)$$

The above equation can be expanded as

$$\sigma = \sigma_1 + \sigma_2 - \frac{\sigma_1 \sigma_2 (\beta_1 - \beta_2)^2 H^2}{\sigma_1 (1 + \beta_2^2 H^2) + \sigma_2 (1 + \beta_1^2 H^2)}, \qquad (8.4)$$

which helps us to lead the final description, Eq. (8.5). In real metals, magnetoresistance defined as $\Delta \rho_{xx} = \rho_{xx}(H) - \rho_{xx}(0)$ is finite and increases with the magnetic field H, where ρ_{xx} is the resistivity tensor. More quantitatively, the magnetoresistance normalized by the transverse resistance is given by

$$\frac{\Delta\rho_{xx}}{\rho_{xx}(0)} = \frac{\sigma_1\sigma_2(\beta_1 - \beta_2)^2 H^2}{(\sigma_1 + \sigma_2)^2 + H^2(\beta_1\sigma_1 + \beta_2\sigma_2)^2}. \tag{8.5}$$

Here, $\rho_{xx} = 1/\sigma$ and $\rho_{xx}(0) = 1/(\sigma_1 + \sigma_2)$. Please remember here that Fermi surfaces composed of both electrons and holes exist in polyvalent metals such as Al, Zn, and Pb [4]. This equation indicates that magnetoresistance is finite except $\beta_1 = \beta_2$ and is proportional to H^2 under low magnetic fields. The ordinary magnetoresistance was first discovered by Thomson for pieces of iron. The magnitude of magnetoresistance for simple metals, however, has been known to be small [5].

Fe-Co and Fe-Ni FM alloys exhibit GMR and a characteristic that is used in magnetic field sensors. Furthermore, the magnitude is changed, depending on the direction of the magnetic field, to that of the electrical current. The linearized Boltzmann transport equation is expressed as

$$\left(-\frac{\partial f_0}{\partial \epsilon}\right) v_k \cdot \left\{-\left(\frac{\epsilon(k) - \zeta}{T}\right)\nabla T + (-e)\left(E - \frac{\nabla\zeta}{(-e)}\right)\right\} =$$
$$-\left(\frac{\partial f}{\partial t}\right)_{\text{scatter}} + v_k\cdot\frac{\partial\phi}{\partial r} + \frac{(-e)}{\hbar}(v_k \times B)\cdot\frac{\partial\phi}{\partial k}.$$

The terms including ∇T, $\nabla\zeta$, and B are diminished, and the first term of the right side becomes ϕ/τ as can be confirmed from Eq. 8.7. Thus, the above equation is further rewritten as

$$\frac{\partial\phi}{\partial z} + \frac{\phi}{\tau v_z} = \frac{eE}{mv_z}\frac{\partial f_0}{\partial v_x}. \tag{8.6}$$

The variation of parameters method brings about the following expression:

$$\phi = \frac{eE\tau}{m}\frac{\partial f_0}{\partial v_x}\left\{1 + A\exp\left(\frac{z}{\tau v_x}\right)\right\}. \tag{8.7}$$

The current density was calculated from v_x and ϕ under the assumption that (a) angular dependence of scattering at the Fe/Cr interfaces can be neglected, (b) there is only transmission or diffusive scattering, and (c) scattering at the outer boundaries is purely diffusive. This model explained all the major features of GRM observed for Fe/Cr layered structures [6].

In addition to the normal magnetoresistance and GMR widely observed, CMR as a further critical phenomenon has been reported

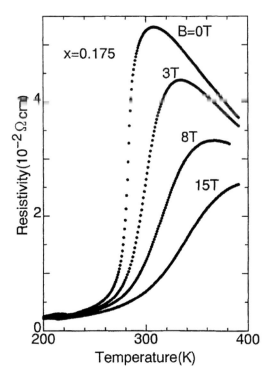

Figure 8.1 Temperature dependence of electrical resistivity under 0, 3, 8, and 15 T for $La_{0.825}Sr_{0.175}MnO_3$ [7].

for manganese oxides. The mechanism of GMR is distinctly different from the magnetoresistive effect observed for metals and/or GMR. CMR occurs by hopping of electrons between the transition metal ions due to the double-exchange interaction on an atomic scale, whereas for GMR materials, the magnetic ordering occurs over tens of Å. Figure 8.1 serves as an example; the electrical resistivity of $La_{0.825}Sr_{0.175}MnO_3$ with the perovskite structure significantly decreases upon an increasing magnetic field, especially at the vicinity of the FM transition temperature (T_c) [7]. This phenomenon can be interpreted as a combination of magnetism and conductivity through the medium of the double-exchange interaction: the FM interaction between the localized spins and conduction electrons.

If one electron with spin s ($= 1/2$ or $-1/2$) hops from site 1 to site 2, where ions possess spin S_1 and S_2, respectively, the binding energy becomes

$$-2J\, \mathbf{S}_1 \cdot \mathbf{s} = -J\left[(\mathbf{S}_1 + \mathbf{s})^2 - \mathbf{S}_1^2 - \mathbf{s}^2\right], \tag{8.8}$$

where J is the intratomic exchange interaction between $\mathbf{s}$ and $\mathbf{S}_1$. Corresponding to the spin function $\alpha\ (= 1/2)$ and $\beta\ (= -1/2)$, the binding energy is expressed as

$$\begin{aligned}\epsilon_1(\alpha) &= -J\left[\left(S + \frac{1}{2}\right)\left(S + \frac{1}{2} + 1\right) - S(S+1) - \frac{1}{2}\left(\frac{1}{2} + 1\right)\right] \\ &= -J\, S \end{aligned} \tag{8.9}$$

and

$$\begin{aligned}\epsilon_1(\beta) &= -J\left[\left(S - \frac{1}{2}\right)\left(S - \frac{1}{2} + 1\right)\right. \\ &\quad \left. - S(S+1) - |-\frac{1}{2}|\left(|-\frac{1}{2}| + 1\right)\right] \\ &= J\,(S+1). \end{aligned} \tag{8.10}$$

Similar procedure for ion 2 leads to

$$\epsilon_2(\alpha') = -J\, S \tag{8.11}$$

and

$$\epsilon_2(\beta') = J\,(S+1). \tag{8.12}$$

Next, let us now attempt to expand the discussion to the situation where the direction of $\mathbf{S}_1$ and $\mathbf{S}_2$ is inclined by θ. Under the favor of the conversion into the spherical coordinates, the Pauli spin matrix $\mathbf{s}$ is transformed to $\mathbf{s}'$:

$$\mathbf{s}' = a\begin{pmatrix}0 & 1\\ 1 & 0\end{pmatrix} + b\begin{pmatrix}0 & -i\\ i & 0\end{pmatrix} + c\begin{pmatrix}1 & 0\\ 0 & -1\end{pmatrix} = \begin{pmatrix}c & a - ib\\ a + ib & -c\end{pmatrix}, \tag{8.13}$$

where $a = \sin\theta\cos\varphi$, $b = \sin\theta\sin\varphi$, and $c = \cos\theta$, and they are connected by the relationship of $a^2 + b^2 + c^2 = 1$ (see Fig. 8.2).

When the wave function $\varphi(\mathbf{s}_z)$ is converted into $\varphi'(\mathbf{s}'_z) = (A, B)$,

$$\begin{pmatrix}c & a - ib\\ a + ib & -c\end{pmatrix}\begin{pmatrix}A\\ B\end{pmatrix} = \lambda\begin{pmatrix}A\\ B\end{pmatrix}, \quad \lambda = \pm 1 \tag{8.14}$$

holds.

Comparison with the first component yields

$$\frac{|A|^2}{|B|^2} = \frac{|a - ib|^2}{|1 - c|^2} = \frac{1 - c^2}{(1 - c)^2} = \frac{\cos^2(\theta/2)}{\sin^2(\theta/2)}, \tag{8.15}$$

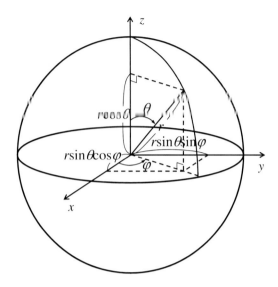

Figure 8.2 Spherical coordinate r, θ, and φ.

in which $a^2 + b^2 + c^2 = 1$ and $c = \cos\theta$ are used for the third and fourth term expansions.

The matrix element of the Hamiltonian is given by

$$
\begin{pmatrix}
-J\,S & 0 & b\cos(\theta/2) & -b\sin(\theta/2) \\
0 & J\,(S+1) & -b\sin(\theta/2) & b\cos(\theta/2) \\
b\cos(\theta/2) & -b\sin(\theta/2) & -J\,S & 0 \\
-b\sin(\theta/2) & b\cos(\theta/2) & 0 & J\,(S+1)
\end{pmatrix},
\qquad (8.16)
$$

where b is the transfer integral between sites 1 and 2.

Its eigenvalue is found to be

$$
\epsilon = \frac{1}{2}J \pm \left[J^2\left(s + \frac{1}{2}\right)^2 + b^2 \pm 2J\,b(s+1)\cos\frac{\theta}{2} \right]^{1/2}. \qquad (8.17)
$$

Accordingly, the degree of transfer is concluded to be determined by θ.

At temperatures below T_c, FM interaction and also magnetic fields become the angle θ zero, resulting in a decrease in ϵ. This is a phenomenological explanation of the negative magnetoresistance near T_c. Extensive research has been devoted to but not limited to manganese oxides. The relatively large magnetoresistance found in cobalt oxide is introduced in the next section.

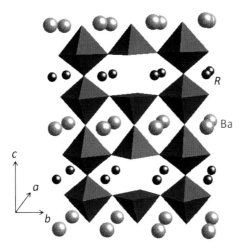

Figure 8.3 Crystal structure of $RBaCo_2O_{5.5}$. The CoO_5 square pyramids and CoO_6 octahedra are shown by the blue color.

8.3 Magnetoresistance of $RBaCo_2O_{5.5}$

Recently, layered cobaltites $RBaCo_2O_{5+x}$ (R: the rare-earth element) with square-lattice CoO_2 planes came into focus because of their remarkable transport and magnetic properties. We will begin by comprehending the crystal structure of the compound. The doubling of the b axis originates from an alternation of CoO_5 square pyramids and CoO_6 octahedra along this direction, while the doubling along the c axis is due to the stacking layer of $[BaO][CoO_2][RO_{0.5}][CoO_2]$ planes (see Fig. 8.3) [8]. A variety of spin and orbital states is available in this system, owing to its rich phase diagram, where $x = 0.5$ is the parent compound with all cobalt ions trivalent.

$RBaCo_2O_{5.5}$ shows a series of phase transitions [9]. The metal–insulator transition takes place around 360 K in the paramagnetic (PM) phase. The PM phase also shows transitions to the FM phase upon cooling and then to antiferromagnetic (AF) phases, which is accompanied by the onset of a strong anisotropic magnetoresistance effect below 255 K. As an origin of these transitions, various contradicting magnetic structures, including different spin states of Co^{3+} ions, have been proposed on the basis of neutron diffraction measurements.

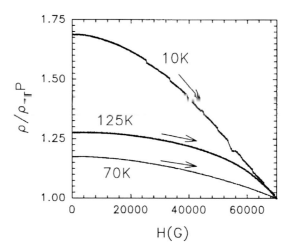

Figure 8.4 Magnetic field dependence of electrical resistivity normalized that measured under 7 T for $GdBaCo_2O_{5.4}$. The data taken at 10 K, 70 K, and 125 K upon heating is plotted [8].

Among several interesting properties, we are concerned with magnetoresistance in this section. The appearance of a giant negative magnetoresistance was demonstrated by Martin et al. for the first time [8]. In their paper, a resistance ratio between under 0 T and 7 T, R_0/R_{7T}, reaches at least 10 at 10 K and is significantly larger than that observed for other cobalt oxides. As shown in Fig. 8.4, the data shows a negative magnetoresistance and the above ratio increases corresponding to the AF transitions, which indicates a link between magnetic and electronic properties. Especially, the ratio is enhanced at lower temperatures. The maximum magnetoresistance percentage, $(\rho_0 - \rho_{7\,T})/\rho_0$, is estimated to be 41% [8].

Historically, Khomskii and Kugel explained the ferromagnetism of K_2CuF_4 in conjunction with AF orbital ordering to make the energy of Hamiltonian [10]. The novel ferromagnetism was also explained from the viewpoint of the orbital-ordering model proposed by Taskin et al. [11]. In their model, $3d$ orbitals bring about the intermediate-spin (IS) state of the Co^{3+} order coupled antiferromagnetically along the a axis at 350 K, which induces ferromagnetism. Around this temperature, thermally excited carriers, between which the double-exchange interaction works, may

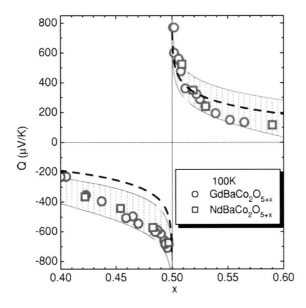

Figure 8.5 Doping variation of thermoelectric power at 100 K with changing oxygen content in GdBaCo$_2$O$_{5+x}$ and NdBaCo$_2$O$_{5+x}$. The dashed lines above and below $x = 0.50$ are the celebrated Heikes formula $Q = (k_B/e)\ln(3/2 - x)/(x - 1/2)$ and $Q = -(k_B/e)\ln(1/2 + x)/(1/2 - x)$, respectively, setting a standard $x = 0.50$. The hatched area represents the entire available range of the entropy contribution [12].

give spins a positive effect on forming FM ordering. These 1D FM spins are aligned antiferromagnetically along the b axis due to the superexchange interaction, resulting in AF transition at 255 K.

Furthermore, the variability of oxygen content in RBaCo$_2$O$_{5+x}$ allows one to dope continuously either electrons or holes in the CoO$_2$ planes. The validity of the electron or hole doping is tested by the oxygen content dependence of thermoelectric power. As displayed in Fig. 8.5, the thermoelectric power diverges around the valence of $3+$ ($x = 0.5$) and its absolute value goes down in the negative- and positive-sign region by electron and hole doping, respectively, [12], which is in good conformity with the theory proposed by Koshibae et al. [18].

8.4 Extremely Large Magnetoresistance of PdCoO$_2$

PdCoO$_2$ crystallizes in the delafossite-type structure which has alternative layers with triangular lattice [13]. Pd triangular layers and CoO$_2$ triangular slabs stack alternatively along the c axis (see Fig. 8.6).

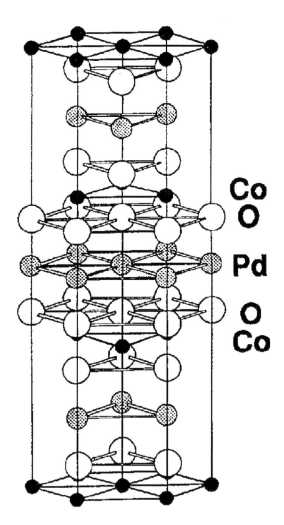

Figure 8.6 Crystal structure of PdCoO$_2$ [14].

Although most of the delafossiate-type oxides are semiconductive antiferromagnets, $PdCoO_2$ shows a metallic conductivity with the anisotropy $\rho_{||}/\rho_{\perp}$ of 149 at 260 K, where $\rho_{||}$ and $\rho_{\perp}$ are the in-plane and out-of-plane resistivity, respectively [14]. The thermoelectric power is also typical value expected from the metallic characteristics, i.e., 2–4 $\mu V/K$ between 100 and 400 K [15]. The electronic structure is investigated by photoemission spectroscopy; the density of states (DOS) in the vicinity of the Fermi level mainly comes from Pd $4d$ electrons [16]. Furthermore, a deep cave looks like a pseudo-gap is formed at the Fermi level. The non-magnetic low-spin Co^{3+} essentially leads to an interpretation of non-magnetic $PdCoO_2$.

We shall concentrate on an extremely large magnetoresistance with a positive response for $PdCoO_2$ [17]. Let me stress here that GMR and CMR ever reported are a negative response and realized by the coupling between spin and charge. $\rho_{\perp}$ is strongly enhanced by applying the magnetic field along $[1\bar{1}0]$ direction, reaching 35000% of resistivity under zero field at 2 K (see Fig. 8.7). When we apply the magnetic field along the [110] direction, $\rho_{\perp}$ increases, but still remains the metallic trend. The azimuthal field angle variation can be captured semiclassical calculations based on tight-binding model, together with the enhancement of $\rho_{\perp}$. Thus, the observed behavior is attributable to the Lorentz force ($\propto v_F \times \boldsymbol{B}$) characterized by the peculiar Fermi surface due to $v_F \propto -\nabla\varepsilon(k_F)$, rather than effects of the spin-orbit interaction.

We can say with fair certainty that a confining effect originates from three factors: (i) the Fermi surface with the hexagonal prism, (ii) two-dimensional layered structure where electrons move freely, and (iii) excellent metallic conductivity.

8.5 Spin Blockade

In an oxygen-deficient perovskite $HoBaCo_2O_{5.5}$, Co^{3+} is normally considered to be in the IS state in the pyramids and in the LS state in the octahedra at lower temperatures, while both in the sites are in the HS state or IS state above the metal–insulator transition temperature (T_{MI}) of ≈ 285 K (see Fig. 8.8). In addition

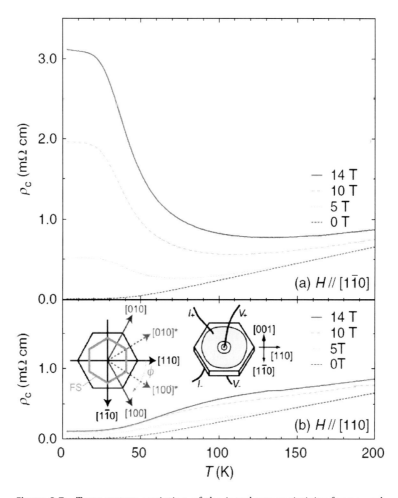

Figure 8.7 Temperature variation of the interlayer resistivity for several magnetic fields along the (a) [1$\bar{1}$0] and (b) [110] directions [17].

to the electrical resistivity variation in the vicinity of T_{MI}, the thermoelectric power changes its sign from positive to negative upon heating. To borrow the scenario proposed by Maignan et al.: the electron excitation or hole excitation with increasing temperature makes the ground state of the trivalent Co ion Co^{2+} or Co^{4+}, respectively, [19]. Judging from the sign of thermoelectric

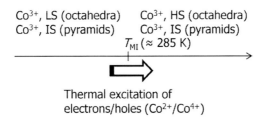

Figure 8.8 Valence and spin state of $HoBaCo_2O_{5.5}$.

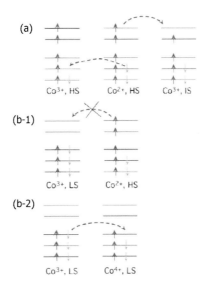

Figure 8.9 Schematic illustration of the process of electron hoping on the background of (a) the HS and IS Co^{3+} and (b) LS Co^{3+}. A spin blockade phenomenon is sketched in (b-1) [19]. *Abbreviations*: HS, high spin; LS, low spin.

power, it is no exaggeration to say that the contributions of Co^{2+} and Co^{4+} dominate above and below T_{MI}, respectively.

Having got information on the valence and spin state, we may now turn to the real subject. We concentrate on the mobility effect of the spin state of the background Co^{3+} on Co^{2+} and Co^{4+}. When Co^{2+} is surrounded by HS or IS Co^{3+} (see Fig. 8.9a), electrons in Co^{2+} can move to their sites, which means a simple exchange of both the spin state and the valence. However, in the case of LS Co^{3+}, the situation is turned around; that is, one cannot move one of the electrons of Co^{2+} to Co^{3+}, retaining the same spin state in contrast to LS Co^{3+}-LS Co^{4+} assortment, as shown in Fig. 8.9b. The process where one cannot interchange spin states by moving an electron is called a spin blockade, resulting in a blockade of the current. Spin blockade generally occurs in a situation in which an electron is unable to access an energetically favorable path, owing to spin conservation [20], and is also the basis of a lot of recent progress in spintronics.

References

1. Thomson W (1856–1857), On the electro-dynamic qualities of metals: effects of magnetization on the electric conductivity of nickel and of iron, *Proc. R. Soc. London*, **8**, 546–550.

2. Binasch G, Grunberg P, Saurenbach F, Zinn W (1989), Enhanced magnetoresistance in layered magnetic structures with antiferromagnetic interlayer exchange, *Phys. Rev. B*, **39**, 4828–4830.

3. Baibich MN, Broto JM, Fert A, Nguyen Van Dau F, Petroff F, Etienne P, Creuzet G, Friederich A, Chazelas J (1988), Giant magnetoresistance of (001)Fe/(001)Cr magnetic superlattices, *Phys. Rev. Lett.*, **61**, 2472–2475.

4. Cracknell AP (1971), *The Fermi Surface of Metals*, Taylor & Francis Ltd.

5. Coleman RV, Morris RC, Sellmyer DJ (1973), Magnetoresistance in iron and cobalt to 150 kOe, *Phys. Rev. B*, **8**, 317–331.

6. Camley RE, Barnas J (1989), Theory of giant magnetoresistance effects in magnetic layered structures with antiferromagnetic coupling, *Phys. Rev. Lett.*, **63**, 664–667.

7. Tokura Y, Urushibara A, Moritomo Y, Arima T, Asamitsu A, Kido G, Furukawa N (1994), Giant magnetotransport phenomena in filling-controlled Kondo lattice system: $La_{1-x}Sr_xMnO_3$, *J. Phys. Soc. Jpn.*, **63**, 3931–3935.

8. Martin C, Maignan A, Pelloquin D, Nguyen N, Raveau B (1997), Magnetoresistance in the oxygen deficient $LnBaCo_2O_{5.4}$ (Ln = Eu, Gd) phases, *Appl. Phys. Lett.*, **71**, 1421–1423.

9. Respaud M, Frontera C, Garcia-Munoz JL, Aranda MAG, Raquet B, Broto JM, Rakoto H, Goiran M, Llobet A, (2001), Magnetic and magnetotransport properties of $GdBaCo_2O_{5+\delta}$: a high magnetic-field study, *Phys. Rev. B*, **64**, 214401-1–214401-7.

10. Kugel KI, Khomskii DI (1982), The Jahn-Teller effect and magnetism: transition metal compounds, *Sov. Phys. Usp.*, **25**, 231–256.

11. Taskin AA, Lavrov AN, Ando Y (2002), Ising-like spin anisotropy and competing antiferromagnetic-ferromagnetic orders in $GdBaCo_2O_{5.5}$ single crystals, *Phys. Rev. Lett.*, **90**, 227201-1–227201-4.

12. Taskin AA, Lavrov AN, Ando Y (2006), Origin of the large thermoelectric power in oxygen-variable $RBaCo_2O_{5+x}$ ($R = Gd, Nd$), *Phys. Rev. B*, **73**, R121101-1–R121101-4.

13. Prewitt CT, Shannon RD, Rogers DB, (1971), Chemistry of noble metal oxides. II. Crystal structures of platinum cobalt dioxide, palladium cobalt dioxide, coppper iron dioxide, and silver iron dioxide, *Inorg. Chem.*, **10**, 719–723.

14. Tanaka M, Hasegawa M, Takei H, (1996), Growth and anisotropic physical properties of $PdCoO_2$ single crystals, *J. Phys. Soc. Jpn.*, **65**, 3973–3977.

15. Hasegawa M, Inagawa I, Tanaka M, Shirotani I, Takei H, (2002), Thermoelectric power of delafossite-type metallic oxide $PdCoO_2$, *Solid State Comm.*, **121**, 203–205.

16. Higuchi T, Tsukamoto T, Tanaka M, Ishii H, Kanai K, Tezuka Y, Shin S, Takei H, (1998), Photoemission study on $PdCoO_2$, *J. Electron Specrosc.*, **92**, 71–75.

17. Takatsu H, Ishikawa J, Yonezawa S, Yoshino H, Shishidou T, Oguchi T, Murata K, Maeno Y, (2013), Extremely large magnetoresistance in the nonmagnetic metal $PdCoO_2$, *Phys. Rev. Lett.*, **111**, 056601-1-056601-5.

18. Koshibae W, Tsutsui K, Maekawa S (2000), Thermopower in cobalt oxides, *Phys. Rev. B*, **62**, 6869–6872.

19. Maignan A, Caignaert V, Raveau B, Khomskii D, Sawatzky G (2004), Thermoelectric power of $HoBaCo_2O_{5.5}$: possible evidence of the spin blockade in cobaltites, *Phys. Rev. Lett.*, **93**, 026401-1–026401-4.

20. Ono K, Austing DG, Tokura Y, Tarucha S (2002), Current rectification by Pauli exclusion in a weakly coupled double quantum dot system, *Science*, **297**, 1313–1317.

Chapter 9

Intrinsic Inhomogeneity

9.1 Prologue

Homogeneity/inhomogeneity has stirred up interest as one of the origins of unusual physical properties, say, a colossal magnetoresistive effect and superconductivity [1]. Taking advantage of first-order transitions may give rise to either a homologous change of volume or a phase separation (segregation). There are two scenarios for understanding the appearance of unusual physical phenomena. One scenario is that a homogeneous phase is formed perfectly in the material. The other scenario is the phase separation among multiphases, which has been claimed to be attributable to the formation of the ordered state. We should not overlook that this type of separation is essentially different from that well observed for the binary-alloy system in the way of intrinsic, or perhaps it would be more concrete to say that the former phase separation could occur even in the perfect crystal due to the strong correlation between electrons [3].

For example, the phase separation into a hole-rich and hole-poor region can be confirmed, especially for $J \ll t$ and $J \gg t$ (see Fig. 9.1). From the experimental side, for example, the inhomogeneity with a nanoscale spatial variation has been pointed out by using

Functional Cobalt Oxides: Fundamentals, Properties, and Applications
Tsuyoshi Takami
Copyright © 2014 Pan Stanford Publishing Pte. Ltd.
ISBN 978-981-4463-32-4 (Hardcover), 978-981-4463-33-1 (eBook)
www.panstanford.com

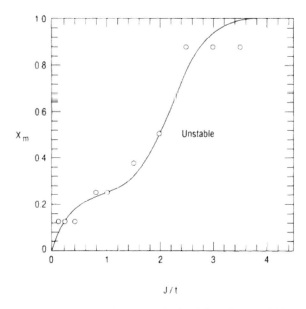

Figure 9.1 J/t variation of x_m, at which $e(x) = (e_H + e_h(x))/x$ takes a minimum, where e_H and e_h are the energy per site in the Heisenberg (hole-free) phase and in the hole-rich phase, respectively [3].

the scanning tunneling microscopy (STM) technique for a high-T_c superconductor $Bi_2Sr_2CaCu_2O_{8+\delta}$ [4, 5]. As is clear from Fig. 9.2, there are different gap Δ regions against homogeneous distribution, which indicates the presence of a superconducting region (smaller Δ), a mixture of two different orders, and a second phase (larger Δ). Turning now to the macroscopic scale, the conclusion that the intrinsic inhomogeneity of carriers may be attributable to disordered and frustrated granular superconductivity is derived from the magnetic irreversibility curve for $YBa_2Cu_4O_8$ having no oxygen vacancy, twin formation, and microscopic disorder [6].

Let's inflate one's image far intuitively by the simple model. High-T_c superconductivity occurs by doping charges to a Mott insulator, in which spins are coupled antiferromagnetically forming an antiferromagnetic (AF) state. The spins are energetically stable by AF coupling by $|J|$. The break of the coupling loses the total energy of $4|J|$ per charge, as shown by $\times$ in Fig. 9.3a. Given the independent motion of two charges, the energy loss becomes $8|J|$,

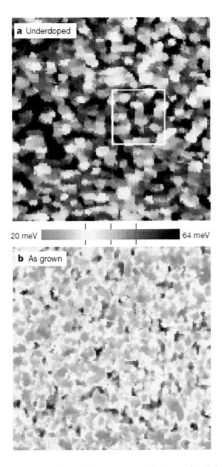

Figure 9.2 Gap map examined by the STM technique for (a) an underdoped and (b) as-grown $Bi_2Sr_2CaCu_2O_{8+x}$ [5].

but it does $7|J|$ by closing together (Fig. 9.3b). To sum up, the energy gain becomes $(N-1)|J|$ for N charges, which indicates a spontaneous phase separation.

Although inhomogeneity arises from the self-organization of electrons, this phenomenon is, of course, not limited to physics, chemistry, and biology. Research on self-organization that extends to sociology, economics, and computer science becomes a major tide as the so-called complex system. This novel phenomenon, being contrary to the physical law that the sum of entropy in

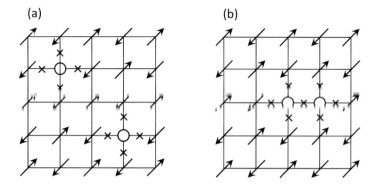

Figure 9.3 Way of charge distribution when each charge is (a) away and (b) near against the backdrop of an AF matrix. The arrow and ○ represent the spin and charge, respectively.

nature continues to increase, is classified into two categories. A static ordering is induced by a large motive force due to the local chemical bonding and only has short-range correlation. In the meanwhile, a long-range correlation can exist in the whole space for a dynamic ordering, and the pattern formation perhaps occurs on the macroscopic scale. These phenomena are called self-assembly and self-organization, respectively.

In the following sections, the spin cluster, the polaron cluster, and phase separations between the metallic and semiconducting phases and between the spin-density wave (SDW) and ferromagnetic phases are introduced as representative phenomena observed for cobalt oxides, which are hardly observed for ordinary materials.

9.2 Spin Cluster

Figure 9.4 shows the magnetic phase diagram of $La_{1-x}Sr_xCoO_3$ [7]. Hole-doped cobaltites $La_{1-x}Sr_xCoO_3$ exhibit spin-glass and superparamagnetic behavior for $0.05 \leq x \leq 0.17$, beyond which they show ferromagnetism coming from a kind of cluster-glass because of an increase in the number and possibly the size of the clusters, together with metallic conduction [7]. For the mother material $x = 0$, as is already well introduced in Chapter 2, it seems reasonable

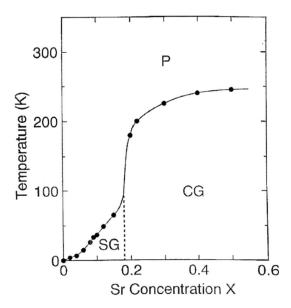

Figure 9.4 Magnetic phase diagram for $La_{1-x}Sr_xCoO_3$. P, SG, and CG represent paramagnetic, spin-glass, and cluster-glass phases, respectively, [7].

to suppose that Co^{3+} undergoes successively a spin-state crossover from the low-spin (LS) state to the intermediate-spin (IS) state around 100 K and finally to the high-spin (HS) state around 500 K. When Sr is doped, the IS state is known to be stabilized, namely, the LS state is diminished, but for the very lightly doped region of $x \le 0.02$, few spins embedded in a nonmagnetic LS matrix give an order of magnitude larger magnetic susceptibility than the expected doped picture [8]. This phenomenon apparently cannot be understood by a homogeneous doping picture.

The Zeeman splitting detected by the inelastic neutron scattering (INS) measurements can be explained with a large effective g factor of 10 [9]. A large number of lines of electron spin resonance (ESR) spectra implies the existence of resonating centers with larger spin multiplicity [9]. For the temperature variation of the nuclear relaxation rate derived from ^{59}Co nuclear magnetic resonance (NMR) measurement, the observed stretch-exponential shape of the nuclear magnetization recovery suggests a substantially nonuniform

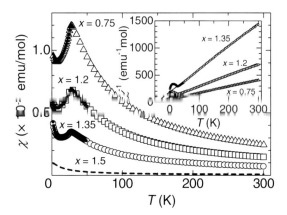

Figure 9.5 Temperature dependence of magnetic susceptibility measured in zero-field-cooling (open symbol and dashed curve) and field-cooling (solid symbol) modes in a field of 10 kOe for $Li_x Co_{0.5} RhO_3$; inset shows the reciprocal magnetic susceptibility except for $x = 1.5$. The solid lines in the inset are the fit to the Curie–Weiss law [11].

distribution of local magnetic environments [9]. Combining with these results, holes doped in the LS state of $LaCoO_3$ transform the six nearest-neighboring Co^{3+} ions to the IS state, forming octahedrally shaped spin-state polarons. Spin clusters behave like magnetic nanoparticles embedded in an insulating nonmagnetic matrix.

9.3 Polaron Cluster

Among Li oxide materials, layered compounds expressed by $Li_2 MO_3$ (M: the transition metal) containing a $Li_{1/3} M_{2/3} O_2$ layer consists of 2D triangular lattices of Li/M ions forming a network of edge-sharing LiO_6/MO_6 octahedra in the ab plane, between which the Li layer exists. Thus, the composition ratio is simply converted into the form $LiLi_{1/3} M_{2/3} O_2$, that is, $Li_2 MO_3$. Recently, $Li_2 RhO_3$ ($M =$ Rh) was synthesized, and its magnetic and transport properties were reported [10]. In our study, the conductive layer is forced to be formed by Rh and its homologous element Co instead of Li. The refinement confirms that our crystal is a single phase with the pseudohexagonal $R\bar{3}m$ symmetry rather than the monoclinic

(a)　　　　　　　　　　(b)

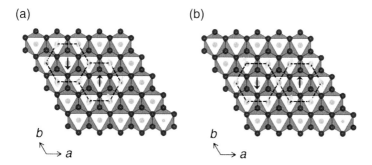

Figure 9.6 Possible configurations of (a) edge- and (b) corner-sharing polarons in the c plane view of $Li_xCo_{0.5}RhO_3$. Red and blue circles are Co/Rh and O atoms, respectively. Red arrows couple antiferromagnetically, and yellow arrows represent induced spins on the six adjacent Rh^{3+} or Co^{3+} [11].

$C2/m$ symmetry of Li_2RhO_3, indicating disorder of Co and Rh in the $Co_{1/3}Rh_{2/3}O_2$ layers. Their magnetic and transport properties are investigated by the author's group [11].

Measurements of magnetic susceptibility indicate that Co^{3+} and Rh^{3+} are both diamagnetic with the LS configuration for $Li_{1.5}Co_{0.5}RhO_3$. The electrical resistivity and thermoelectric power reveal 1D variable-range hopping (1D-VRH) below 320 K (T_1) and above 630 K. Charges introduced by Li extraction down to $x = 0.75$ in $Li_xCo_{0.5}RhO_3$ yield systematic decreases in resistivity and T_1, but an AF ordering is observed at $T_N = 30$ K (see Fig. 9.5). Further attention should be paid to the fact that AF transition already appears for 10% substitution. For instance, no long-range ordering has been revealed down to $x = 0.7$ in Li_xCoO_2 with the 2D triangular lattice consisting only of Co, which is also a band insulator for $x = 1$ [12, 13]. For a 3D array, when electrons are induced in $SrTiO_3$ with no d electron, which is likewise assigned to a band insulator, the paramagnetic–AF transition as well as the insulator–metal transition occur for $x \geq 0.9$ in $Sr_{1-x}La_xTiO_3$ [14].

Spins induced by the removal of 10% of Li ($x = 1.35$) in an LS matrix, for instance, have an effective magnetic moment approximately six-fold larger than that assuming a homogeneous HS Rh^{4+} distribution, which can be interpreted by transforming of involved Co^{3+} or Rh^{3+} ions to the magnetic states. A random, short-range distribution of such ordered clusters would support $R\bar{3}m$ symmetry as against the monoclinic $C2/m$ symmetry of long-range-ordered Li_2RhO_3. Disordered Co and Rh in the $Co_{1/3}Rh_{2/3}O_2$ layers would result in randomly ordered clusters illustrated in Fig. 9.6, in which edge- and corner-sharing polarons are formed. Moreover, such clusters would trap holes into half filled th^{11} orbitals forming bonds to neighbors within the ordered clusters.

Transport between clusters could be well described by VRH and mixed-valent transport within a cluster consisting of edge- and/or corner-sharing polarons with 1D alignment via a single Rh-4d orbital, which would be a possible origin of temperature dependence identical to that in 1D transport. Since geometrical frustration in the 2D configuration of the triangular lattice prevents long-range ordering, this interpretation agrees with the susceptibility data with the AF transition.

9.4 Phase Separation

9.4.1 Metallic and Semiconducting Phases

Figure 9.7 shows the crystal structure of a new compound $Sr_4Co_2O_6CO_3$ containing C, being isostructural with $Sr_4Fe_2O_6CO_3$ [15], which is a derivative $(n = 3)$ of the Ruddllesden–Popper $A_{n+1}Co_nO_{3n+1}$ (A: the alkaline-earth metal), with the space group $I4/mmm$. To be specific, one of three Co atoms is deficient and all the oxygen sites are occupied by O and C atoms with a ratio of 9:1.

Figure 9.8 shows plots of the in-plane and out-of-plane resistivity for the $Sr_4Co_2O_6CO_3$ single crystal grown by the flux method [16]. ρ showed an anisotropic behavior, as is speculated from the crystal structure. In particular, the in-plane resistivity (ρ_{ab}) increases upon cooling and levels off below 80 K. It is important to bear in mind that this behavior is in contrast to the typical metallic or semiconducting conductivity.

The core of the mechanism of the unconventional $\rho_{ab}(T)$ behavior below 80 K is a parallel resistor model, in which when temperature is decreased, the conduction of the semiconducting part $\rho_s(T)$ expressed as $\rho_s[\exp(E_s/k_BT) + 1]$ freezes, and then a

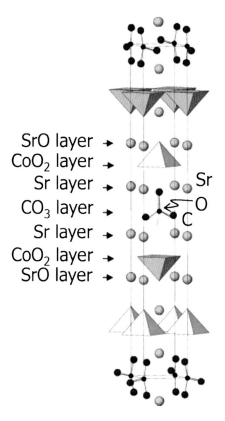

SrO layer →
CoO$_2$ layer →
Sr layer →
CO$_3$ layer →
Sr layer →
CoO$_2$ layer →
SrO layer →

Figure 9.7 Crystal structure of Sr$_4$Co$_2$O$_6$CO$_3$. Co and O atoms in the CoO$_4$ pyramid are passed over.

metallic part $\rho_0 + \alpha T$ plays a dominant role in the $\rho(T)$ curve, given by

$$\rho(T) = \frac{1}{(\rho_0 + \alpha T)^{-1} + \rho_s(T)^{-1}} \tag{9.1}$$

where ρ_s, E_s, k_B, ρ_0, and α are the proportional coefficient, the energy gap, the Boltzmann constant, the residual resistivity, and the temperature coefficient, respectively, as reported in the normal-state resistivity for the superconductors Ba$_{1-x}$K$_x$BiO$_3$ [17]. In the first place, to verify this scenario, we tried to fit Eq. 9.1 to the data in the whole temperature range investigated but failed for full fitting. Our best fit by Eq. 9.1 was completed up to 150 K, and

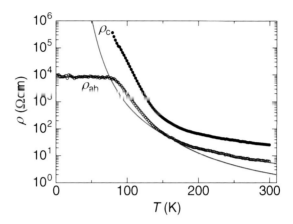

Figure 9.8 Temperature dependence of electrical resistivity of Sr_4 $Co_2O_6CO_3$. ρ_{ab} and ρ_c represent the in-plane and out-of-plane resistivity, respectively. The solid curves with blue and red color show the results of fitting the activation-type model and the parallel resistor to the data, respectively [16].

the measured data deviated upward above this temperature. The $\rho_{ab}(T)$ curve above 150 K, on the other hand, is understood by a simple activation-type conduction, expressed as $\rho = \rho_0 \exp(\Delta / k_B T)$ with a gap energy Δ of 68 meV, up to at least 300 K. These results indicate two conducting channels that are made up of a metallic phase and a semiconducting phase; especially the electronic structure corresponding to the latter phase possesses the different gap energies at a boundary of 150 K. Judging both from the resistivity and the thermoelectric power data, the homogeneous semiconducting phase is modified to the intrinsic inhomogeneous phase, that is, the mixed phase with the metallic and semiconducting regions perhaps via a spin-state crossover from the HS to the LS state of Co^{2+} around 150 K upon cooling the remaining Co^{3+}, the nonmagnetic LS state, as is evidenced by the susceptibility data. More noteworthy is that the finite value of the electronic specific heat coefficient, 9.8 mJK^2 mol^{-1}, also supports the existence of the metallic phase at low temperature.

A parallel response to the electrical field is actually realized for $Sr_4Co_2O_6CO_3$, by which we would be a step closer to new guideline

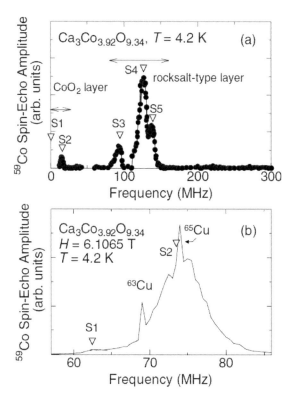

Figure 9.9 Frequency-swept ^{59}Co NMR spectra for $Ca_3Co_{3.92}O_{9.34}$ at 4.2 K under (a) a zero field and (b) 6.1065 T. S1–S5 represent the peak positions of the NMR spectra. The peaks observed at 68.91 and 73.82 MHz in (b) are the ^{63}Cu and ^{65}Cu NMR signals in an NMR coil, respectively [19].

beyond the framework of the conventional homogeneous picture for designing functional materials, if the same response combines with other external perturbations. As an example, under the condition that phase A possesses 100 μV/K (S_A) and phase B possesses −95 μV/K (S_B), the thermoelectric power S of a whole system simply becomes $S = 1/(S_A^{-1} + S_B^{-1}) \approx -1900$ μV/K, which means crucial improvement of the power factor S^2/ρ and the dimensionless figure of merit ZT, given no alternation of ρ and κ. One can safely state that the guideline based on controlling the way of response is effective to the physical quantities with both positive and negative signs.

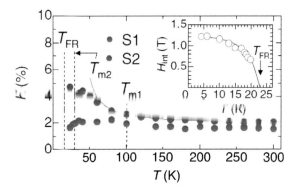

Figure 9.10 Temperature dependences of Knight shift and internal fields of the NMR signals (S1 and S2) for $Ca_3Co_{3.92}O_{9.34}$. T_{m1}, T_{m2}, and T_{FR} are the onset temperatures of a short-range order of an IC SDW state, a long-range IC SDW state, and the ferrimagnetic transition, respectively. The solid curve in the main panel is a fit of the Curie–Weiss law [19]. *Abbreviation*: IC, incommensurate.

9.4.2 *Spin-Density Wave and Ferromagnetic Phases*

Misfit layered cobalt oxides are promising candidates as thermoelectric materials because of their large thermoelectric power, low electrical resistivity, and low thermal conductivity, as in the case of Na_xCoO_2. In particular, they exhibit excellent thermoelectric performance at high temperature, since they are more stable at higher temperatures than Na_xCoO_2. In their lattice, there are at least two Co sites; that is, in the CoO_2 layer and in the rock-salt layer. The charge carrier transport is restricted mainly to the CoO_2 layer, which means that the transport properties are mostly governed by carriers in this layer. Interestingly, the degeneracy of spins and orbitals of the 3d electrons of the Co ions has been theoretically pointed out to be important to enhance the thermoelectric power [18]. However, the local magnetic properties in each layer have not been fully established on the grounds that both the layers include magnetic Co ions. The NMR method offers the key to an understanding of this issue.

NMR is a phenomenon in which an atomic nucleus in the magnetic field absorbs/radiates electromagnetic waves with a specific frequency determined by the nuclide and the magnetic field.

The atomic nucleus has a spin angular momentum of $\hbar\mathbf{I}$, and the magnetic moment $\boldsymbol{\mu}$ is expressed as

$$\boldsymbol{\mu} = g_N\mu_N\mathbf{I} = \gamma_N\hbar\mathbf{I}, \tag{9.2}$$

where g_N, μ_N, and γ_N are the g factor of nucleus, the nucleus magneton, and the nucleus gyromagnetic ratio, respectively. The energy of magnetic moment under the static field $\mathbf{H}_0$ is given as follows:

$$\mathcal{H}_z = -\boldsymbol{\mu}\cdot\mathbf{H}_0 = -\gamma_N\hbar\mathbf{I}\cdot\mathbf{H}_0. \tag{9.3}$$

The eigenstate for the nucleus spin under the magnetic field is written as

$$E = -\gamma_N\hbar H_0 m \quad (m = -I, -I+1, \cdots, I-1, I), \tag{9.4}$$

having $(2I + 1)$ energy levels at $\gamma_N\hbar H_0$ intervals.

To pave the way for transitions between the Zeeman levels, one can apply the oscillating magnetic field along the x direction, say, $\mathbf{H}_1(t) = 2H_1\cos\omega t\,\mathbf{i}$, where $\mathbf{i}$ is the unit vector. The total Hamiltonian becomes

$$\mathcal{H} = -\mathbf{H}\cdot\boldsymbol{\mu} \tag{9.5}$$
$$= -\mathbf{H}_0\cdot\boldsymbol{\mu} - \mathbf{H}_1(t)\cdot\boldsymbol{\mu} \tag{9.6}$$
$$= \gamma\hbar H_0 I_z - \gamma\hbar 2H_1 I_x\cos\omega t. \tag{9.7}$$

Since the amplitude of the oscillating magnetic field is generally much smaller than H_0, the second term can be dealt with as the perturbation. Applying the relationship of , where $I^{+/-}$ is the raising/lowering operation, the matrix element of the perturbation Hamiltonian is derived to be

$$V|I_z > \propto I_x|I_z >= (I^+ + I^-)/2|I_z > \tag{9.8}$$
$$\propto A|I_{z+1} > +B|I_{z-1} > . \tag{9.9}$$

Therefore, the transition between level m and levels $m\pm 1$ occurs when the condition of $\omega = \gamma_N H_0$ is fulfilled. The interaction that affects NMR is called the hyperfine interaction between the nucleus spin and the electron, which is classified into electrical and magnetic ones. The former includes the electric quadrupole interaction, while the latter does the magnetic dipole interaction, the nucleus spin orbital interaction, the Fermi contact interaction, and the interaction by inner shell polarization.

The main advantages of the NMR technique are as follows: NMR can (i) prove information about the low-energy spin motion of electrons due to a quite small nucleus magnetic moment and (ii) extract the local electronic structure of a specific element since the atomic nucleus depends on the element. For other purposes, NMR is available to investigate the structure of complex materials with a large molecular weight, for example, proteins, in chemistry and biology and is also applicable to magnetic resonance imaging (MRI) in medical science.

Figure 9.9 shows frequency-swept ^{59}Co NMR spectra for $Ca_3Co_{3.92}O_{9.34}$ at 4.2 K under (a) a zero field and (b) 6.1065 T [19]. It has been reported that Co in the rock-salt-type layer is divalent and that in the CoO_2 layers is in a mixed valency of trivalent/tetravalent [20]. The number of Co sites that are affected by the misfit is larger in the rock-salt-type layer [21]. In light of the situation, the NMR spectra (S3–S5) at higher frequencies are attributed to the Co in the rock-salt layer, whereas those (S1 and S2) at lower frequencies to the Co in the CoO_2 layer. Although the NMR signal coming from Co with lower internal fields is hard to detect under zero field, measurements under a finite magnetic field overcome the issue. As shown in Fig. 9.9b, S1 and S2 are actually confirmed to exist.

Let us devote a little more space to examining the S1 and S2 signals. Their changes are displayed as the ^{59}Co Knight shift K and internal fields H_{int} in Fig. 9.10. Here, $K = (f_r - f_0)/f_0$ and $H_{int} = 2\pi(f_r - f_0)/\gamma$, where $f_0 = \gamma H/2\pi$ with $\gamma = 2\pi \times 10.054$ MHz/T and $H = 6.1065$ T. K for S1 was about 1.8%, while K for S2 showed a temperature dependence and obeyed the Curie–Weiss law above T_{m1}, then showed a plateau at T_{m2}, and finally H_{int} increased significantly below T_{FR}, where T_{m1}, T_{m2}, and T_{FR} are the onset temperatures of a short-range order of an IC SDW state, a long-range IC SDW state, and the ferrimagnetic transition, respectively [22, 23]. The NMR spectrum corresponding to S1 is accounted for by the NMR shape function in the SDW ordered state. In our NMR experiments, thus, the coexistence of the SDW and FM orders is proved. The ground state for the CoO_2 layer in $Ca_3Co_{3.92}O_{9.34-\delta}$ is summarized with the phase diagram shown in Fig. 9.11. We can be fairly certain

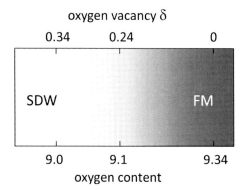

Figure 9.11 Magnetic phase diagram realized in the CoO_2 layer for $Ca_3Co_{3.92}O_{9.34-\delta}$ ($\delta = 0$, 0.24, and 0.34). SDW and FM are the spin-density wave ordered state and the ferromagnetic ordered state, respectively [19].

that the phase-separated state is formed in the CoO_2 layer despite consisting of a crystallographically independent single Co site and the volume fraction of both phases systematically changed with the oxygen content, that is, the carrier concentration.

References

1. Goodenough JB, Zhou JS (1997), New forms of phase segregation, *Nature*, **386**, 229–230.

2. Pan SH, OfNeal JP, Badzey RL, Chamon C, Ding H, Engelbrecht JR, Wang Z, Eisaki H, Uchida S, Gupta AK, Ng KW, Hudson EW, Lang KM, Davis JC (2001), Electronic inhomogeneity in the high-T_c superconductor $Bi_2Sr_2CaCu_2O_{8+x}$, *Nature*, **413**, 282–285.

3. Emery VJ, Kivelson SA, Lin HQ (1990), Phase separation in the *t-J* model, *Phys. Rev. Lett.*, **64**, 475–478.

4. Pan SH, O'Neal JP, Badzey RL, Chamon C, Ding H, Engelbrecht JR, Wang Z, Eisaki H, Uchida S, Guptak AK, Ngk KW, Hudson EW, Lang KM, Davis JC (2001), Microscopic electronic inhomogeneity in the high-T_c superconductor $Bi_2Sr_2CaCu_2O_{8+x}$, *Nature*, **413**, 282–.

5. Lang KM, Madhavan V, Hoffman JE, Hudson EW, Eisaki H, Uchida S, Davis JC (2002), Imaging the granular structure of high-T_c superconductivity in underdoped $Bi_2Sr_2CaCu_2O_{8+\delta}$, *Nature*, **415**, 412–416.

6. Takami T, Shimokata A, Itoh M (2010), Growth of $YBa_2(Cu,Co)_4O_8$ single crystals under ambient pressure and their superconducting properties, *J. Phys. Soc. Jpn.*, **79**, 014711-1–014711-4.

7. Itoh M, Natori I, Kubota S, Motoya K (1994), Spin-glass behavior and magnetic phase diagram of $La_{1-x}Sr_xCoO_3$ ($0 \leq x \leq 0.5$) studied by magnetization measurements, *J. Phys. Soc. Jpn.*, **63**, 1486–1493.

8. Yamaguchi S, Okimoto Y, Taniguchi H, Tokura Y (1996), Spin-state transition and high-spin polarons in $LaCoO_3$, *Phys. Rev. B*, **53**, R2926–R2929.

9. Podlesnyak A, Russina M, Furrer A, Alfonsov A, Vavilova E, Kataev V, BuNchner B, StraNssle T, Pomjakushina E, Conder K, Khomskii DI (2008), Spin-state polarons in lightly-hole-doped $LaCoO_3$, *Phys. Rev. Lett.*, **101**, 247603-1–247603-4.

10. Todorova V, Jansen M (2011), Synthesis, structural characterization and physical properties of a new member of ternary lithium layered compounds: Li_2RhO_3, *Z. Anorg. Allg. Chem*, **637**, 37–40.

11. Takami T, Cheng JG, Goodenough JB (2012), Magnetic and transport properties of layered $Li_xCo_{0.5}RhO_3$, *Appl. Phys. Lett.*, **101**, 102409-1–102409-3.

12. Sugiyama J, Nozaki H, Brewer JH, Ansaldo EJ, Morris GD, Delmas C (2005), Frustrated magnetism in the two-dimensional triangular lattice of Li_xCoO_2, *Phys. Rev B*, **72**, 144424-1–144424-9.

13. Motohashi T, Ono T, Sugimoto Y, Masubuchi Y, Kikkawa S, Kanno R, Karppinen M, Yamauchi H (2009), Electronic phase diagram of the layered cobalt oxide system Li_xCoO_2 ($0.0 \leq x \leq 1.0$), *Phys. Rev B*, **80**, 165114-1–165114-9.

14. Hayo CC, Zhou JS, Market JT, Goodenough JB (1999), Electronic transition in $La_{1-x}Sr_xTiO_3$, *Phys. Rev. B*, **60**, 10367–10373.

15. Yamaura K, Huang Q, Lynn JW, Erwin RW, Cava RJ (2000), Synthesis, crystal structure, and magnetic order of the layered iron oxycarbonate $Sr_4Fe_2O_6CO_3$, *J. Solid State Chem*, **152**, 374–380.

16. Takami T, Tsuchihashi K, Kawano R (2011), Contribution of carriers in both the metallic and semiconducting phase to the transport properties in the Sr-Co-O-C system, *Appl. Phys. Lett.*, **99**, 072102-1–072102-3.

17. Nagata Y, Mishiro A, Uchida T, Ohtsuka M, Samata H (1999), Normal-state transport properties of $Ba_{1-x}K_xBiO_3$ crystals, *J. Phys. Chem. Solids*, **60**, 1933–1942.

18. Koshibae W, Tsutsui K, Maekawa S (2000), Thermopower in cobalt oxides, *Phys. Rev. B*, **62**, 6869–6872.

19. Takami T, Nanba H, Umeshima Y, Itoh M, Nozaki H, Itahara H, Sugiyama J (2010), Phase separation in the CoO_2 layer observed in thermoelectric layered cobalt dioxides, *Phys. Rev. B*, **81**, 014401-1–014401-12.

20. Masset AC, Michel C, Maignan A, Hervieu M, Toulemonde O, Studer F, Raveau B (2000), Misfit-layered cobaltite with an anisotropic giant magnetoresistance: $Ca_3Co_4O_9$, *Phys. Rev. B*, **62**, 166–175.

21. Miyazaki Y, Onoda M, Oku T, Kikuchi M, Ishii Y, Ono Y, Morii Y, Kajitani T (2002), Modulated structure of the thermoelectric compound $[Ca_2CoO_3]_{0.62}CoO_2$, *J. Phys. Soc. Jpn.*, **71**, 491–497.

22. Sugiyama J, Xia C, Tani T (2003), Anisotropic magnetic properties of $Ca_3Co_4O_9$: evidence for a spin-density-wave transition at 27 K, *Phys. Rev. B*, **67**, 104410-1–104410-5.

23. Sugiyama J, Brewer JH, Ansaldo EJ, Itahara H, Dohmae K, Seno Y, Xia C, Tani T (2003), Hidden magnetic transitions in the thermoelectric layered cobaltite $[Ca_2CoO_3]_{0.62}[CoO_2]$, *Phys. Rev. B*, **68**, 134423-1–134423-8.

Chapter 10

Move/Diffuse and Charge/Discharge Effect

10.1 Prologue

A key word for the Internet revolution was "concentration to dispersion" at the end of the 20th century; that is, a large-scaled computer that dominates everything was converted into a personal computer or workstation. Recent developments of technology transmogrified a computer with high performance, for example, to a portable phone, in the same way as the dispersion launched in the field of energy, in which fuel cells have an important role. Fuel cells, which are a generation device made by using the electrochemical reaction between a fuel (H_2) and air (O_2), are divided into four cells on the basis of the kind of electrolytic solution: polymer electrolyte fuel cell (PEFC), phosphoric acid fuel cell (PAFC), molten carbonate fuel cell (MCFC), and solid oxide fuel cell (SOFC). Their characteristics are summarized in Table 10.1. MCFC is currently being developed and marketed, whereas PEFC that can be easily operated at low temperature and SOFC with high efficiency have also attracted attention. In particular, a solid oxide from fuel cells is an ultimate power generator, which can expect the highest

Functional Cobalt Oxides: Fundamentals, Properties, and Applications
Tsuyoshi Takami
Copyright © 2014 Pan Stanford Publishing Pte. Ltd.
ISBN 978-981-4463-32-4 (Hardcover), 978-981-4463-33-1 (eBook)
www.panstanford.com

Table 10.1 Characteristics of various fuel cells

	PAFC	PEFC	SOFC	MCFC
Electrode	phosphoric acid	solid polymer	ZrO	carbonate
Fuel	natural gas, methanol	H_2, natural gas	natural gas, LPG	natural gas, LPG
Operating temperature (°C)	200	80–100	700–1000	600–700
Efficiency (%)	40–45	40–60	45–65	45–65
Characteristics	emergency power supply	household use	household use	distributed medium-scale

power generation efficiency among power generators, and since it is able to collect waste of heat simultaneously, its technology has attracted attention as a new energy supply system that contributes to realization of a low-carbon society, in addition to the promotion of energy saving. Section 10.2 deals with SOFC consisting of Co oxides as a cathode material.

There is one more topic that comes within the scope of this chapter. Hydrogen, the most abundant element in the universe, has great potential as an energy source. Among metals, it is known that there are several hydrogen storage materials, especially Mg alloys and V alloys. For compatible absorption and desorption, space in the crystal structure to absorb hydrogen is required and the hydrogen atom exists stably to some extent at the position, yet hydrogen has to be capable to move/diffuse. Hazardous materials such as HC, CO, and NO_x can be removed by using the catalyst device composed of Pt, Pd, and Rh. The oxygen storage and release properties, on the other hand, have become one of the important functions of present automotive emission control. In this chapter, these two topics are discussed from the viewpoint of oxygen vacancy of Co oxides.

10.2 Cathode Material of Solid Oxide Fuel Cell: $Sr_{0.7}Y_{0.3}CoO_{2.63}$

For SOFC, hydrogen and/or neutral gas is normally used as a fuel at the operating temperatures of 923–1273 K and the conduction ion is O^{2-}. The structure of SOFC would look something like that in Fig. 10.1. The O atom receiving the electron at the cathode becomes an O^{2-} ion, then it is carried to the electrolyte and arrives at the anode, and finally the electron is released after combining with H_2. The chemical reactions occur at the cathode and the anode, respectively:

$$1/2O_2 + 2e^- \rightarrow O^{2-} \tag{10.1}$$

and

$$H_2 + O^{2-} \rightarrow H_2O + 2e^-. \tag{10.2}$$

It follows from what has been said that we can regard the electrodes where oxidation and reduction take place as an anode and a cathode,

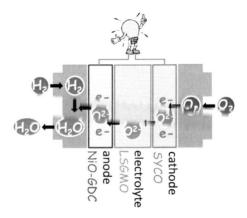

Figure 10.1 Illustration of the structure of SOFC. SYCO, LSGMO, and GDC represent $Sr_{0.7}Y_{0.3}CoO_{2.63}$, $La_{0.8}Sr_{0.2}Ga_{0.83}Mg_{0.17}O_{2.815}$, and Gd-doped ceria, respectively.

respectively. The above two chemical formulae amount to saying that the reaction as a battery is written as

$$H_2 + 1/2O_2 \rightarrow H_2O. \tag{10.3}$$

Suitable conditions for the cathode material of SOFC are as follows: thermodynamical stability, good electrical and ionic conductivity, affinity for oxygen, and poor reaction with other compositional materials.

SOFC converts the chemical energy of the reaction of a fuel with oxygen directly into electrical energy. The rate-limiting step of this electrochemical conversion is the oxygen–reduction reaction at the cathode. A cathode active for the oxygen–reduction reaction must be a good electronic conductor, but delivery of the electrons from the anode to the adsorbed O_2 on the cathode may occur either near a triple-phase boundary between the air/cathode/O^{2-} ion electrolyte or over a broader surface area of a mixed O^{2-} ion/electron conductor in contact with the O^{2-}-ion solid electrolyte. In this section, let's focus our attention on $Sr_{0.7}Y_{0.3}CoO_{2.63}$ as a good candidate for the cathode material [1].

The crystal structure of $Sr_{0.7}Y_{0.3}CoO_{2.65-\delta}$ [2] is shown in Fig. 10.2. The symmetry is the tetragonal $I4/mmm$ space group. There are crystallographically two kinds of Co sites and four oxygen

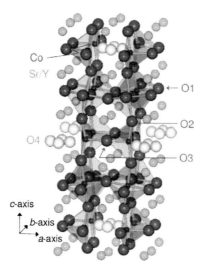

Figure 10.2 Crystal structure of $Sr_{0.7}Y_{0.3}CoO_{2.63}$.

sites. The high electronic conductivity and the large amount of disordered oxygen vacancies, as well as the presence of loosely bound O4 atoms, motivated us to evaluate the electrochemical performance as a potential cathode material. This system is a mixed O^{2-}-ion/electron conductor exhibiting good activity for the oxygen–reduction reaction, which makes it a competitive cathode material.

Figure 10.3 shows the temperature dependences of electrical resistivity in a logarithmic scale (left) and conductivity in a linear

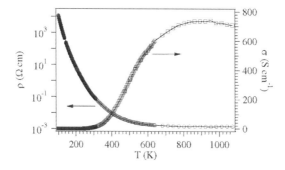

Figure 10.3 Temperature dependences of electrical resistivity in a semilog scale (left) and conductivity in a linear scale (right) for $Sr_{0.7}Y_{0.3}CoO_{2.63}$ [2].

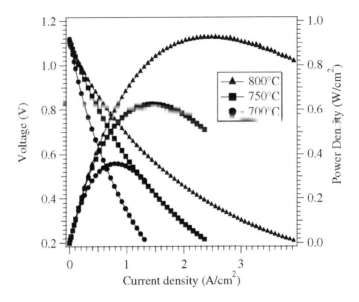

Figure 10.4 Cell voltage (left) and power density (right) as a function of current density for the SYCO|LSGM|NiO-GDC test cell at different temperatures [2].

scale (right) over a wide temperature range from 100 K to 1073 K. With all, the sample shows a semiconducting behavior at low temperature, first decreases sharply by 6 orders up to 400 K with increasing temperature, then it exhibits a much weaker temperature dependence, and eventually it becomes metallic above 900 K. The value of conductivity satisfies the general requirement for the cathode materials of SOFC, namely, 100 $\Omega^{-1}cm^{-1}$ above 400 K, especially reaching 735 $\Omega^{-1}cm^{-1}$ around 920 K.

The cell is actually composed of a cathode, an anode, and an electrode, as I mentioned at the outset; the electrode performance was first evaluated by electrochemical impedance spectroscopy on a symmetrical cell SYCO|LSGM|SYCO, where SYCO and LSGM are abbreviated forms of $Sr_{0.7}Y_{0.3}CoO_{2.63}$ and $La_{0.8}Sr_{0.2}Ga_{0.83}Mg_{0.17}O_{2.815}$, respectively. A SYCO cathode with an area of 0.25 cm^2 was screen-printed onto both sides of a dense LSGM electrolyte pellet of 600 μm thickness. The impedance response for the oxygen reduction on SYCO at 700, 750, and 800 °C in air is colored as a depressed arc.

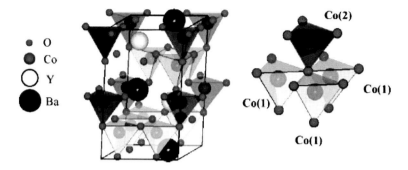

Figure 10.5 Crystal structure of YBaCo$_4$O$_7$ [5].

The high-frequency intercept of the electrode impedance on the real axis is the total resistance of the electrolyte. The difference between the low-frequency and high-frequency intercepts on the real axis corresponds to the area-specific resistance of the two interfaces. The area-specific resistance is the overall resistance related to the oxygen reduction, the oxygen surface/bulk diffusion, and the gas-phase oxygen diffusion. The area-specific resistance values and the total resistance reduce significantly with increasing temperature, the former of which reaches 0.11 Ωcm^2 at 800 °C. This indicates that SYCO has high electrocatalytic activity for the oxygen–reduction reaction at intermediate temperatures.

Figure 10.4 depicts the cell voltage and power density as a function of current density for the single fuel cell NiO-GDC|LSGM|SYCO operating at different temperatures with dry hydrogen as the fuel and ambient air as the oxidant, where GDC is Gd-doped ceria. The voltage decreased monotonically as the current density is increased, but the power density, that is, the product of voltage and current density, increases and takes a maximum value at each temperature. The maximum power density P_{max} reaches 0.36, 0.63, and 0.93 W/cm^2 at 700, 750, and 800 °C, respectively, nearly double the practical requirements of 0.5 W/cm^2 at 800 °C for a single cell. It follows from such an excellent cell performance that SYCO is a promising cathode material for SOFCs. For example, (Ba,Sr)(Co,Fe)O$_{3-\delta}$ and LaBaCo$_2$O$_{5+\delta}$ exhibit 1.0 W/cm^2 at 554 °C and 0.52 W/cm^2 at 800 °C, respectively [3, 4].

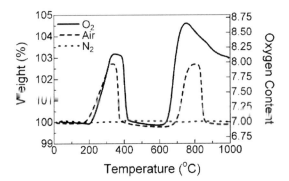

Figure 10.6 TG curve measured in O_2, air, and N_2 atmospheres for $YBaCo_4O_7$ [5].

10.3 Oxygen Storage Material: $YBaCo_4O_{7+\delta}$

Let's expand the argument into the second topic. Besides the fact that the presence of oxygen vacancy has oxygen ions moving/diffusing in the crystal lattice, some materials reversibly charge/discharge oxygen. Oxygen-nonstoichiometric $YBaCo_4O_{7+\delta}$ is known to possess large low-temperature oxygen absorption/desorption capability [5]. Its crystal structure is shown in Fig. 10.5; two kinds of corner-sharing CoO_4 tetrahedral, $Co(1)O_4$ and $Co(2)O_4$, with a ratio of 3:1 exist. Figure 10.6 shows the thermogravimetric (TG) curve measured in O_2, air, and N_2 atmospheres. In particular, TG curves except N_2 exhibit pronounced humps at two temperature ranges around 200–400 °C and 600–900 °C, but the second one is due to the decomposition of $YBaCo_4O_{7+\delta}$. The former essential change yields 1880–2440 μ-O/g, being higher than that of CeO_2-ZrO_2, (1500 μ-O/g [6]) works on the basis of the reversible redox reaction between Ce^{4+} and Ce^{3+}, which proves its value as an oxygen storage material.

In the research for gas-related materials, the storage/release of oxygen is also realized for the multiferroic $LuFe_2O_{4+x}$ (0 < *x* < 0.5) reflecting its structural flexibility [7]. Oxyhydrides are relatively rare, in which the hydride anion (H^-) replaces oxygen, but $LaSrCoO_3H_{0.7}$ and $Sr_3Co_2O_{4.33}H_{0.84}$ with extremely low oxidation states, i.e., $Co^{1.7+}$ and $Co^{1.75+}$, respectively, have been reported [8, 9].

For another transition-metal oxide, $BaTiO_{3-x}H_x$ is exchangeable with hydrogen gas [10].

References

1. Takami T, Cheng JG (2011), Effects of structural disorder and charge carriers on the magnetic and transport properties of $Sr_{0.7}R_{0.3}CoO_{3-\delta}$ ($R = Y$ or $Dy_{0.45}Er_{0.55}$), *Jpn. J. Appl. Phys.*, **50**, 013002-1–013002-5.

2. Li Y, Kim YN, Cheng JG, Alonso JA, Hu Z, Chin YY, Takami T, Fernández-Díaz MT, Lin HJ, Chen CT, Tjeng LH, Manthiram A, Goodenough JB (2011), Oxygen-deficient perovskite $Sr_{0.7}Y_{0.3}CoO_{2.65-\delta}$ as a cathode for intermediate-temperature solid oxide fuel cells, *Chem. Mater.*, **23**, 5037–5044.

3. Tarancon A, Skinner SJ, Chater RJ, Hernandez-Ramirez F, Kilner JA (2007), Layered perovskites as promising cathodes for intermediate temperature solid oxide fuel cells, *J. Mater. Chem.*, **17**, 3175–3181.

4. Kim JH, Manthiram A (2008), $LnBaCo_2O_{5+\delta}$ oxides as cathodes for intermediate-temperature solid oxide fuel cells, *J. Electrochem. Soc.*, **155**, B385–B390.

5. Karppinen M, Yamauchi H, Otani S, Fujita T, Motohashi T, Huang YH, Valkeapaa M, Fjellvåg H (2006), Oxygen nonstoichiometry in $YBaCo_4O_{7+\delta}$: large low-temperature oxygen absorption/desorption capability, *Chem. Mater.*, **18**, 490–494.

6. Nagai Y, Yamamoto T, Tanaka T, Yoshida S, Nonaka T, Okamoto T, Suda A, Sugiura M (2002), X-ray absorption fine structure analysis of local structure of CeO_2-ZrO_2 mixed oxides with the same composition ratio ($Ce/Zr = 1$), *Catal. Today*, **74**, 225–234.

7. Hervieu M, Guesdon A, Bourgeois J, Elkaïm E, Poienar M, Damay F, Rouquette J, Maignan A, Martin C (2013), Oxygen storage capacity and structural flexibility of $LuFe_2O_{4+x}$ ($0 < x < 0.5$), *Nature Mater.*, **13**, 74–80.

8. Hayward MA, Cussen EJ, Claridge JB, Bieringer M, Rosseinsky MJ, Kiely CJ, Blundell SJ, Marshall IM, Pratt FL (2002), The hydride anion in an extended transition metal oxide array: $LaSrCoO_3H_{0.7}$, *Science*, **295**, 1882–1884.

9. Helps RM, Rees NH, Hayward MA (2010), $Sr_3Co_2O_{4.33}H_{0.84}$: An extended transition metal oxide-hydride, *Inorg. Chem.*, **49**, 11062–11068.

10. Kobayashi Y, Hernandez OJ, Sakaguchi T, Yajima T, Roisnel T, Tsujimoto Y, Morita M, Noda Y, Mogami Y, Kitada A, Ohkura M, Hosokawa S, Li Z, Hayashi K, Kusano Y, Kim J, Tsuji N, Fujiwara A, Matsushita Y, Yoshimura K, Takegoshi K, Inoue M, Takano M, Kageyama H (2012), An oxyhydride of $BaTiO_3$ exhibiting hydride exchange and electronic conductivity, *Nature Mater.*, **11**, 507–511.

Index

air 155, 160, 162
alloys 75, 109, 157
Anderson transition 40
anode 47, 50–51, 53, 157–58, 160
antiferromagnetic transition,
    partially disordered 88, 92,
    96, 100, 104
antiferromagnetism 78, 85
antiferromagnitic coupling 82, 88,
    138
antiferromagnitic order 88, 104–5
antiferromagnitic transition
    128–29, 143–44
atomic nucleus 1, 148–50

band insulator 143
batteries 45–47, 50–51, 53, 58, 72,
    158
    biological 45
    chemical 45
    history of 46–47
    lithium ion 45–46, 48, 50, 52
    physical 45
    rechargeable 46–47
    rechargeable lithium ion 45–46
    wet cell 47
BCS (Bardeen, Cooper, and
    Schrieffer) theory 109, 111
Bloch–Wilson transition 40
Bose–Einstein Condensation
    110–11

cathode 47, 51, 53, 157–58, 160
cathode material 48, 157–58,
    160–61
cell voltage 160–61
charge-ordered insulating states
    71
chemical reactions 45, 51, 157
clusters
    octahedral 10–11
    ordered 144
    polaron 140, 142–43
Co ions 10, 27–29, 32, 36, 40–41,
    65, 67, 82, 93, 99, 103–4, 148
    spin state of 27–29, 67, 99
    trivalent 30, 35
    valence and spin state of 28–29
Co oxides 1, 27, 45, 63–64, 69,
    78–79, 115, 126, 128, 140,
    157
conduction electron system 65, 67
conductivity 36, 114, 122, 124,
    159–60
CoO_2 27, 35, 53, 61–63, 71, 127,
    148, 150–51
CoO_5 127
CoO_6 10–11, 82
CoO_6 octahedra 87–88, 92–93
    edge-sharing 48, 61, 63, 88
CoO_6 trigonal prism 87, 92–93
copper 46–47
copper oxides 116
corner-sharing polarons 143–44
Coulomb energy 11–12
crystal field 9–10, 30

crystal field parameters 29
crystal field splitting 5, 29–30
crystal structure 16, 32, 35,
   48–49, 61, 63–65, 79–80, 85,
   87–88, 92–93, 105, 127, 144,
   157–59, 162
crystals, single 62, 89, 144
cuprate superconductors,
   high-T_c 110, 112–13
Curie temperature 76, 114
Curie–Weiss law 32, 37, 98, 104,
   142, 148, 150
Curie–Weiss metal 115
current density 12, 17, 19, 49,
   122–23, 160–61

Daniell cell 46–47
Debye model 21–22
Debye temperature 16, 20
degeneracy 9, 18, 28, 30, 69, 148
depolarizer 47

electric field 5, 9, 122
electrical resistivity 20, 32, 37, 39,
   58–59, 61–63, 69–70, 109,
   121, 124, 128, 143, 146, 148,
   159
electricity 46–48, 53
electrode materials 1
electrodes 46, 51, 156–57, 160
   negative 46, 51
   positive 46–49, 51
electrolyte 47–48, 157, 161
electrolyte solution 46
electrolytic solution 51, 155
electron configuration 9–10
electron distribution 13–14
electron spin resonance (ESR) 141
electron system 1, 67, 86
electron systems, strongly
   correlated 3, 39, 64
electron theory, solid-state 78

electron thermal conductivity 21
electronic states 10, 13
   extended 30
electronic structure 11–12, 34,
   39–40, 53–54, 70, 100, 131,
   146, 150
electrons
   conduction 1, 124
   core 100
   transition 9
   valence 102
elements, rare-earth 27, 32, 35,
   88, 114, 121, 127
energy
   binding 53, 100, 102, 125
   charge transfer 11–12
   crystal field 30
   electrical 45–46, 158
   internal 20–21, 65, 67, 86
entropy 17–18, 69, 139
ESR *see* electron spin resonance

Fe 2, 10, 52, 61, 75–77, 80, 114,
   161
Fe ions 32
Fermi level 27, 31, 39–40, 70, 131
Fermi particles 110
Fermi surfaces 13, 114, 118, 123,
   131
Fermi–Dirac distribution
   function 14, 16, 19
Fermi–Dirac function 13, 65
ferromagnet,
   room-temperature 79
ferromagnetic interaction 91, 124,
   126
ferromagnetic metals 76
ferromagnetic order 81, 99, 104–5
ferromagnetic phases 71, 140, 148
ferromagnetic temperature 115
ferromagnetic transition
   temperature 78, 124

ferromagnetism 75–76, 78–81, 85, 99, 115, 128, 140
  bulk 80–81
  room-temperature 75–76, 78–81
  theory of 78
ferromagnetism in transition metal oxides 78
free-electron model 58, 67
fuel cells 45, 72, 155

gap energies 23, 146
giant magnetoresistance (GMR) 121, 123–24, 131
GMR *see* giant magnetoresistance
graphene 49–50
graphite 50–51
ground state 30, 115, 132, 150
Grove cell 46–47

high spin 28, 82, 104, 133, 141, 146
high spin state 28–30, 32, 82, 131
high-T_c superconductivity 112–13, 138
hole doping 49, 129
hydride anion 162
hydrogen 157

inelastic neutron scattering (INS) 117, 141
inhomogeneity 137, 139
  intrinsic 137–38, 140, 142, 144, 146, 148, 150
INS *see* inelastic neutron scattering
insulators 39–40, 62
interaction
  double-exchange 78–80, 124
  nearest-neighbor 90, 92
intercalation 116
intermediate-spin states 82

inverse susceptibility 96, 98

Jahn–Teller distortion 34–35
Jahn–Teller effect 10, 30

lattice constant 30–31
$LiCoO_2$ batteries 48–49, 51, 53
low spin 28–30, 32, 34–36, 68, 103–4, 133, 141
low-spin state 29–30, 34–35, 67–69, 131, 141
  nonmagnetic 32, 146

magnetic Co ions 32, 148
magnetic field 13, 31, 69, 75, 78, 81, 86, 88–89, 94, 96, 122–23, 126, 131–32, 148–49
  critical 111, 113
  oscillating 149
magnetic interactions 85, 100
magnetic properties 36, 49, 78–79, 86–87, 127
magnetic resonance imaging (MRI) 113, 150
magnetic spins 77
magnetic states 143
magnetic structure 85, 105
magnetic susceptibility 32, 35–36, 88–89, 116–17, 142–43
magnetic transition 36, 85–86, 88
magnetism 75, 78, 86–87, 89–93, 95, 97, 99, 101, 103, 105, 114, 124
magnetization 76–78, 80–81, 89
magnetoresistance 121–23, 127–29
  large 121–22, 126, 131
  negative 126, 128
magnetoresistance effect 122–23, 125
magnetoresistive effects 121, 124

magnets 75
manganese oxides 121, 124, 126
MCFC *see* molten carbonate fuel
cell
metallic states 11
metals 14, 20, 39, 46, 60, 62–63,
65, 109, 121, 123–24, 157
microscopic techniques 99
molecular orbital energy level
11
molten carbonate fuel cell
(MCFC) 155–56
Mott insulator 17, 114, 116, 138
MRI *see* magnetic resonance
imaging

$NaCo_2O_4$ 41, 60–63, 69
crystal structure of 61,
88
Na_xCoO_2 61, 63, 115–18, 148
neutrinos 93
NMR *see* nuclear magnetic
resonance
normalized asymmetry 95–96
nuclear charge 53–54
nuclear magnetic resonance
(NMR) 32, 71, 94, 117, 141,
148–50
nucleus spin 149

octahedral sites 79–80, 99, 104
one-electron band structure
calculations 34
orbital angular momentum 3, 75
orbital degeneracy 41–42
orbital ordering 82
orbital splitting 5, 30
orbital states 127
orbitals 1–3, 5, 7, 9–11, 28–31, 34,
38, 40, 121–22, 124, 126, 128,
130, 132, 134
antibonding 10, 30

atomic 10
molecular 10
orbitals split 30
oxygen 10–11, 32, 105, 111, 158,
162
oxygen reduction 160–61
oxygen reduction reaction 158,
161
oxygen storage material 162
oxygen vacancies 80, 82, 138, 157,
162
oxyhydrides 162

PAFC *see* phosphoric acid fuel cell
paramagnetic metallic state 71
partially disordered
antiferromagnetic
transition 85–86, 88, 90, 92,
94, 96, 98–100, 102, 104
PEFC *see* polymer electrolyte fuel
cell
Peierls transition 40
perovskite structure 10, 32–33,
78, 124
perturbations, external 1, 12–13,
15, 17, 19, 21, 23, 31, 147
phase separation 137, 140,
144–45, 147, 149
phase transitions 39, 127
phases, electronic 3
phonons 20–23
phosphoric acid fuel cell (PAFC)
155–56
physics, solid-state 110
Planck distribution function 20,
22
Poggendorff cell 46–47
polymer electrolyte fuel cell (PEFC)
155–56
prismatic sites 103–4
pudding mold band 70

resistance 161
resistivity 62, 71, 131, 143, 146
  out-of-plane 131, 144, 146
resonating valence band (RVB) 110
rock-salt layer 112, 148, 150
rock-salt-type layer 150
RVB *see* resonating valence band

scanning tunneling microscopy (STM) 138
SDW *see* spin-density wave
SDW state 148, 150
semiconducting phases 140, 144, 146
Slater's law 53–54
SOFC *see* solid oxide fuel cell
  structure of 157–58
solid oxide fuel cell (SOFC) 1, 155–61
spin angular momentum 3, 9–10, 149
spin blockade 121–22, 131, 133–34
spin clusters 140–42
spin configurations 1, 9–10, 28–29, 31, 68
spin-density wave (SDW) 140, 150–51
spin excitation 23
spin gap 23, 39
spin networks 23
spin-state crossover 27–28, 30–38, 40–42, 141, 146
spin-state distributions 37
spin-state polarons 142
spin-state transition 31–32, 35–36, 38–39, 67
spin thermal conductivity 22–23
STM *see* scanning tunneling microscopy

Sugano–Tanabe diagram 29
sulfuric acid 47
superconducting transition temperature 113–14
superconductivity 27, 78, 109–10, 112, 114–16, 137
  discovery of 114
  room temperature 50
superconductors 61, 113, 115, 117
  type II 117

temperature dependence of electrical resistivity 32, 159
temperature dependence of electrical resistivity 146, 159
temperature dependence of thermal conductivity 23, 41
thermal conductivity 13, 19–20, 23–24, 28, 40–41, 58–59, 62, 89
  electronic 19–20, 41
  phonon 22, 41
thermal rectifier 40–41
thermoelectric devices 57
thermoelectric generators 71–72
thermoelectric materials 1, 57–63
thermoelectric power 13, 15, 17–18, 27, 37, 49, 57–60, 62–72, 115, 129, 131, 143, 147–48
thermoelectric technology 71
transition metal ions 51, 79–80, 124
transition metal oxides 32, 78, 110
transition temperatures 34–35
  metal–insulator 131
  spin-state 34–35
transitions
  ferrimagnetic 148, 150
  metal-insulator 39–40

trimers 37–38
  octahedral 36, 38

voltaic pile 46–47

Wiedemann–Franz law 20

X-ray photoemission spectroscopy
  (XPS) 100–1

XPS *see* X-ray photoemission
  spectroscopy

Zaanen–Sawatzky–Allen
  diagram 11, 13
zero field 96–97, 131, 147,
  150
zero temperature, absolute 11, 67,
  109
zinc 47